COMPUTER PROBLEMS for MODERN PHYSICS

Charles W. Leming

Henderson State University

SAUNDERS COLLEGE PUBLISHING

Philadelphia New York Chicago
San Francisco Montreal Toronto
London Sydney Tokyo

Copyright © 1990, by Saunders College Publishing, a subsidiary of Holt, Rinehart and Winston, Inc.

All rights reserved. No part of this publication may be reproduced or transmitted in any form or by any means; electronic or mechanical, including photocopy, recording, or any information storage and retrieval system, without permission in writing from the publisher.

Requests for permission to make copies of any part of the work should be mailed to: Permissions, Holt, Rinehart and Winston, 111 Fifth Avenue, New York, New York 10003.

Cover Credit: Relativistic doppler effect, computer generated by Charles W. Leming.

Printed in the United States of America

COMPUTER PROBLEMS IN MODERN PHYSICS

0-03-046207-X

Library of Congress Catalog Card Number: 89-81178

9012 059 987654321

PREFACE

Physicists and engineers were aware of potentially powerful applications of numerical methods long before computers became available to make these methods usable for other than specialized technical calculations. Widespread practical applications required the ability to perform rapid, repetitive calculations. The first electronic digital computers solved this problem, and the power of numerical methods was exploited from the time of their earliest operation.

The development of computer graphics is more recent. The first graphic display driven by a digital computer consisted of a cathode ray tube (CRT) controlled by MIT's Whirlwind I computer in 1950. By 1962, the first practical interactive graphical interface (called *Sketchpad: A Man-Machine Graphical Communication Interface*) had been developed at MIT.

Even though representational graphics had been used throughout history to create descriptions of physical phenomena and mathematical relationships, similar uses of nonrepresentational graphics are much more recent, dating back only about 200 years. The development of interactive graphic interfaces using color and shading to combine representational and nonrepresentational graphics has permanently and irreversibly changed the ways that physical concepts are expressed. Before graphic computers became widely available, many graphical and numerical techniques were impractical because the necessary combination of artistic and mathematical skills was almost never available. The synthesis of interactive computer graphics and numerical methods has yielded a powerful new tool that literally expands the imagination.

The purpose of this textbook is to apply a combination of elementary numerical and graphical methods to the study of modern physics at the intermediate undergraduate level. Involvement and activity are stressed so that students gain both new skills and improved understanding of modern physics.

Complete program listings in BASIC are included in the body of the text. Students can easily apply the programs using whatever computer is available. The programs are written so that little or no computer programming experience is required to understand the procedures. Students who are familiar with more structured methods of programming or with other programming languages should be encouraged to convert the programs to the idiom that they prefer.

The development of the programs and exercises of this text was guided by classroom experience with help and inspiration from students in the physics classes that I teach at Henderson State University. Reactions of users and reviewers of my earlier text *Computer Problems for Classical Dynamics* (Harcourt Brace Jovanovich, 1988) provided additional guidance.

I am especially grateful to my colleagues Donald Avery, Bryan D. Palmer and John Gentry for help, suggestions and encouragement. Jeff Holtmeier, Kathy Walker, Sarah Randall, and Suzanne Montazer from Harcourt Brace Jovanovich supplied professional expertise. John Risley, North Carolina State University, and Paul Zitzewitz, University of Michigan, Dearborn, provided valuable reviews. My greatest debt is to my wife Paula Leming for technical assistance and for much-needed moral support.

Equipment used in the development of this text was supplied in part by a grant from the College Science Instrumentation and Laboratory Equipment Program of the National Science Foundation. The project was also aided by a summer sabbatical provided by Henderson State University.

Charles W. Leming

Contents

Chapter 1: Introduction

1.1	Problems and Solutions	1
1.2	To the Instructor	3
1.3	To the Student	4
1.4	References	5

Chapter 2: Transformations of Coordinates

2.1	Introduction	6
2.2	Rotation of Coordinates	7
2.3	Translation of Coordinates	10
2.4	Three-Dimensional Transformations	13
2.5	Galilean Transformations	15
2.6	The Classical Doppler Effect	19
2.7	Exercises	22

Chapter 3: Relativistic Transformations

3.1	Introduction	25
3.2	The Lorentz Transformation	26
3.3	Geometric Appearance at Relativistic Speeds	29
3.4	Relativistic Kinematics	33
3.5	The Relativistic Doppler Effect	36
3.6	Relativistic Dynamics	38
3.7	Exercises	42

Chapter 4: Motion in Classical Mechanics

4.1	Introduction	43
4.2	Euler's Method	44
4.3	Last-Point Approximation	46
4.4	Motion of an Oil Droplet	47
4.5	Radioactive Decay	51
4.6	Deflection of Electrons in an Electric Field	53
4.7	Deflection of Electrons in a Uniform Magnetic Field	56
4.8	The Rutherford Scattering Experiment	59
4.9	Exercises	63

Chapter 5: Schrödinger's Equation

- 5.1 Introduction 65
- 5.2 Schrödinger's Equation and the Last-Point Approximation 67
- 5.3 One-Dimensional Square Well Potential 68
- 5.4 The Harmonic Oscillator 73
- 5.5 The Zero Potential 76
- 5.6 The Step Potential 80
- 5.7 The Barrier Potential 83
- 5.8 Fourier Transforms and the Uncertainty Principle 87
- 5.9 Exercises 94

Appendix A: Computer Graphics

- A.1 Introduction 99
- A.2 Screen Concepts 100
- A.3 Graphics Operations 100
- A.4 Graphic Template 101
- A.5 Graphing a Mathematical Function (Cartesian) 105
- A.6 Graphing a Mathematical Function (Polar) 107
- A.7 Kinematics in Two Dimensions 110
- A.8 Travelling Waves 113
- A.9 Exercises 117

Appendix B: Program Conversions

- B.1 Introduction 120
- B.2 Applesoft BASIC 121
- B.3 DEC ReGIS 121

Chapter 1

INTRODUCTION

1.1 Problems and Solutions

This text is intended for students who are using computers to enhance their understanding of the ideas of modern physics. Even though topics are limited to ideas of fundamental importance, this approach allows students to gain insights and study phenomena at a level well beyond that of a descriptive treatment.

This text is intended primarily as a supplement to be integrated with an introductory course in modern physics in which a conventional text is used. If preferred, the text can instead be used as the basis for a mini-course in computational physics with modern physics as a prerequisite or corequisite. The study of modern physics follows the one-year college-level general physics sequence for students in a wide range of disciplines. The materials are thus intended for use by students who have completed general physics and are beginning intermediate-level college physics courses. Students using this text develop practical problem-solving skills while enhancing their understanding of the principles and ideas of modern physics. This method is useful not only for students majoring in physics but is especially valuable for students planning careers in engineering, chemistry, molecular biology, or secondary education.

The computer programs and explanatory materials can be used with any standard modern physics text. Appropriate computer methods are developed and applied in the context of the familiar topics of modern physics. Each section contains a brief explanation of the fundamental principles being explored.

The computer programs make extensive use of graphics. The programs are written in the BASIC programming language and can be adapted for use with almost any computer having graphics capability. *These programs are meant to be run—not merely studied.* For this purpose, complete listings of all programs (including redundant templates) are included in the body of the text. The programs can be used without alteration with an MS-DOS computer using BASICA and CGA, EGA, or VGA graphic adaptors. If other types of computers are available or if other programming languages are preferred, the programs can easily be converted to other formats.

The exercises and programs of the text begin with the topics of classical and relativistic coordinate transformations. Students gain experience with familiar and intuitive classical phenomena and then contrast these with the mind-expanding properties of space and time according to the principles of the special theory of relativity.

Following the study of relativity, historically important experiments leading to the development of the ideas of quantum physics are explored by computer modelling of the motion of particles. These computer programs introduce students to numerical methods of solving differential equations. The programs are based on the ideas of classical physics. However, two distinct goals are achieved by the study of historic experiments. First of all, students can more easily learn new techniques in the context of classical physics. Secondly, students acquire an understanding and appreciation of the importance of the empirical methods used in the development of physics.

Once a solid understanding of these elementary numerical techniques has been developed, students are ready to study the properties of matter on the atomic scale using numerical solutions of Schrödinger's equation. Students apply techniques which have already been developed in previous sections. In this way, students can concentrate on the interpretation and understanding of results rather than computer programming and numerical methods. Students investigate the origins of fascinating and subtle quantum mechanical phenomena such as tunneling, energy quantization, and the uncertainty principle.

This method allows students to experiment with ideas that may otherwise remain understood only as mathematical abstractions rather than as properties of the "real" world. Although the ideas of modern physics must be developed from a theoretical point of view, with the help of a computer, students can conduct numerical "experiments" which they devise for themselves. Students thus create their own understanding of the abstract ideas of relativity and quantum mechanics. By developing skill with the techniques of computer graphics, the results of experimentation are displayed graphically, allowing rapid interpretation and understanding of the results. This procedure combines the best features of the controversial "constructionist" ideas about learning with the traditional directed methods of teaching and learning.

The use of computers and numerical methods has blurred the distinction between theoretical physics and experimental physics. Theoretical ideas are now routinely subjected to numerical experiments. Many physicists even feel that a new branch of physics—called computational physics—has emerged. The techniques of computational physics are yielding new understanding of nature and physical laws. These methods are not just tools for the professional physicist. Physics students can apply the methods of computational physics to enhance their understanding while gaining access to these important tools of the professional physicist.

1.2 To the Instructor

This text is not meant to be a text on computational physics; it is intended as a supplement to enhance a traditional course in modern physics. Many decisions and compromises were necessary in the development of the text. Only a few of these decisions—such as the use of graphical displays and the "laboratory" approach to teaching and learning—are of fundamental importance. Many other decisions are arbitrary and could be altered to fit the local situation.

Arbitrary decisions include the use of the BASIC programming language. This language is simple and accessible while lending itself to experimentation. In a related decision, numerical methods were restricted to elementary techniques. For example, Euler's method and the last-point approximation generate reasonably accurate solutions to ordinary differential equations and can be easily understood by students who have completed introductory calculus.

All of the programs of the text use SI units. This allows students to gain concrete experience with physical quantities on the atomic scale. A practical alternative is to use reduced variables or atomic units. This technique can be useful for advanced students, but beginning students find the results more realistic and satisfying when figures are scaled in terms of the standard SI system of units.

By avoiding reduced variables, other problems were created. All versions of BASIC restrict the magnitude of numbers used in calculations. The numbers encountered in quantum mechanics can easily exceed these limits and produce errors. All of the programs assume that the least capable versions of BASIC have been used and should produce no such errors. However, to avoid these problems—and to speed execution of programs—any of the compiled versions of BASIC recently developed for use with microcomputers are highly recommended.

Once decisions about computer languages and computational methods were made, further decisions on topics and exercises followed. The topics covered are typical of those found in traditional modern physics texts. Analytical results can be found to compare with the results of computer calculations. Extensions of these problems for which analytical solutions often cannot be found are included in the exercises at the ends of chapters. The topics covered can be integrated into a traditional modern physics course. If all of the supplemental materials are used, a few traditional topics might be deleted because of time restrictions.

Instead of integrating the materials of this text into a traditional course, a computer course with modern physics as a prerequisite or corequisite can be offered. An instructor might wish to try this approach before attempting the more difficult problem of integrating computer methods into an already existing modern physics course. In addition, students who completed modern physics at an earlier time might wish to apply computers both to enhance their understanding of modern physics and to learn useful computational and graphical skills.

A list of the topics covered in each chapter is included in the table of contents. Topics can be covered in the order that they appear in a traditional modern physics course. A brief discussion of the BASIC programming language could be included at the beginning of the course. In practice, this is seldom necessary, as even the uninitiated can learn elementary programming techniques by example and without formal instruction.

By contrast, formal instruction in the techniques of computer graphics may be necessary. The exercises of Appendix A can be used to develop familiarity with graphical techniques. Although any computer with graphics capability can be used for these exercises, computers with low graphics resolution may prove to be unsatisfactory. Because graphics statements are not standardized, the graphics syntax may vary with the type of computer or version of the BASIC programming language.

Before the computational methods and their associated computer programs are developed, a separate introduction providing at least descriptive information about the particular physical principles being studied should always be furnished. Introducing new concepts together with unfamiliar methods is often a serious (but tempting) pedagogical error. The computational methods used in the programs of this text are purposely limited to a manageable few in order that computational methods can always be introduced in the context of familiar ideas.

Students should be encouraged to experiment with computer programs in a directed "laboratory" setting. Exercises are included at the end of each chapter to allow students to test their understanding. Most of the exercises can be completed by altering a few lines of the programs of the text. When this approach is used, students will invariably create original extensions and enhancements of the programs. Even though it may often appear frivolous, involvement of this type should be encouraged. These activities create an atmosphere which encourages both learning and communication. Students should be encouraged to explain their results to others in both a formal seminar setting and in the informal environment of the laboratory.

Complete listings of all programs are included in the text. The programs are related as closely as possible to the equations and principles being applied. Any of the programs could be altered for greater computational efficiency and for greater speed and accuracy.

The programs of the text should be used to provide "hands-on" experience for students rather than mere classroom "demonstrations." Using personal computers, students can experiment with systems which they invent and alter at will. Understanding is tested when systems are changed and the results of calculations are tested against intuitive predictions.

1.3 To the Student

The topics of modern physics deal with the behavior of matter, energy, space, and time in the realm of speeds and dimensions outside the bounds of experience. Everyday activities yield no analogies to provide a foundation for understanding ideas such as Einstein's special theory of relativity and Heisenberg's uncertainty principle. Computer programs based on these ideas can be used as a substitute for everyday experience by providing a "laboratory" in which predictions can be made and ideas tested.

The programs of this text are not intended to be used as a means of imparting facts to passive learners. The programs of this text are intended instead to promote active involvement, with the goal of producing an understanding and a knowledge of the ideas and concepts of modern physics. To gain this advantage, students should experiment with the programs and test the limits of the ideas being studied. Unlike results derived using conventional methods, a computer program is not an end in itself.

Computer programs only provide a starting point. For instance, students should extend the programs by adding their own personal touches such as color graphics, sound, and captions. When possible students should feel free to alter the programs and adapt them to work with their own personal computer, as the programs of this text are written using the simplest possible programming statements and will run on a wide variety of computers with only minor modifications. These activities produce involvement and students can examine for themselves the ideas that provide the foundations of the programs.

The computer programs of this text will provide both questions and answers. The exercises suggested at the ends of the chapters can be used to provide topics for investigation. On the other hand, there is no right or wrong way to use the programs. Try out your own variations and ideas. Not only will you enhance your computer skills and problem solving ability, you will create understanding.

1.4 References

A partial list of references of special interest is given below.

W. Newman and R. Sproll, *Principles of Interactive Computer Graphics* (McGraw-Hill, New York, 1979)

S. Koonin, *Computational Physics* (Benjamin/Cummings, Menlo Park, CA 1986)

W. Thompson, *Computing in Applied Science* (John Wiley, New York, 1984)

R. Tufte, *The Visual Display of Quantitative Information* (Graphics Press, Cheshire, CT, 1983)

C. Leming, *Computer Problems for Classical Dynamics* (Harcourt Brace Jovanovich, San Diego, CA, 1988)

M. Burns, *Modern Physics for Science and Engineering* (Harcourt Brace Jovanovich, San Diego, CA 1988)

J. McGervey, *Introduction to Modern Physics* (Academic Press, New York, 1983)

R. Eisberg and R. Resnick, *Quantum Physics* (John Wiley, New York, 1985)

H. Ohanian, *Modern Physics* (Prentice-Hall, Englewood Cliffs, NJ, 1987)

A. Beiser, *Concepts of Modern Physics* (McGraw-Hill, New York, 1987)

R. Weidener and R. Sells, *Elementary Modern Physics* (Allyn and Bacon, Boston, 1982)

R. Serway, C. Moses, and C. Moyer, *Modern Physics* (Saunders, Philadelphia, 1989)

Chapter 2

Transformations of Coordinates

2.1 Introduction

Nature has revealed no preference for any particular coordinate system. This simple idea—called the classical principle of relativity—is a basic element of both classical physics and modern physics. However, unlike nature itself, mathematical descriptions of natural phenomena require specific coordinate systems. The symmetry of the system being described determines which type of coordinate system is best for mathematical descriptions. For example, circularly symmetric systems (such as the vibrational modes of a circular drumhead or the force acting on an orbiting planet) are conveniently described in polar coordinates. Equations describing systems involving uniform fields (such as the motion of a projectile not too far from the earth's surface or the action of the electric field between capacitor plates) are simplest when cartesian coordinates are used.

In addition to the type of coordinate system, the location of the origin of the coordinate system and the orientation of the coordinate axes must also be specified. Because of the classical principle of relativity, the location and orientation of coordinate axes are arbitrary—natural laws are the same in any nonaccelerated coordinate system. We are free to extend and stimulate our imagination by looking at physics from many different points of view, each of which is equally valid.

2.2 Rotation of Coordinates

Only the orientation of coordinate axes is altered when the coordinate system is rotated. In Figure 2.1(a), the point P has coordinates (x,y) with respect to the coordinate system of the figure. However, in Figure 2.1(b), the same point P has new coordinates (x',y') with respect to rotated coordinate axes. The two sets of axes share a common origin of coordinates O, but the axes of Figure 2.1(b) are rotated relative to the axes of Figure 2.1(b).

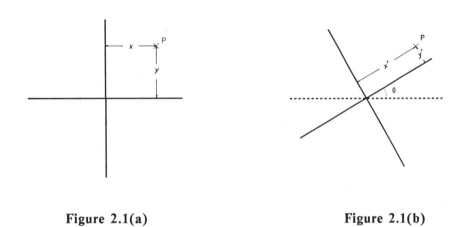

Figure 2.1(a)   Figure 2.1(b)

The coordinate axes of both systems are shown together in Figure 2.2. From the geometry of the figure, transformation equations can be derived to relate (x,y) (the coordinates of P with respect to the axes x and y) to (x',y') (the coordinates of P with respect to the axes x' and y') in terms of θ, the angle between the axes. (See Exercise 1.1.)

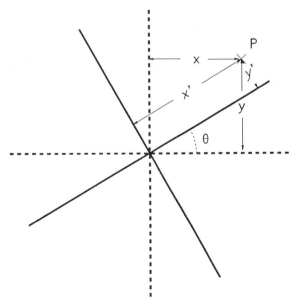

Figure 2.2

2 · TRANSFORMATIONS OF COORDINATES

$$x = x' \cos \theta - y' \sin \theta \tag{2.1}$$

$$y = x' \sin \theta + y' \cos \theta \tag{2.2}$$

With these transformation equations, figures can be rotated relative to the computer screen. The coordinates of the points that form the figure are labelled (x,y) relative to the coordinate axes of the graphic template drawn on the computer screen. Values of x' and y' are determined by the coordinates of the points that make up the figure plotted relative to axes rotated through an angle θ with respect to the axes of the graphic template.

To rotate a figure relative to the computer screen, the coordinates of each point of the figure are determined and the transformation equations are then applied. Plotting the set of points corresponding to these new coordinate locations creates a figure unaltered in shape but drawn with respect to a rotated coordinate system. Program 2.1 uses this technique to plot a rotated version of the cardioid function plotted using Program A.3 (See Appendix A.) Equations 2.1 and 2.2 are applied in line number 3000 and line number 3010, respectively. The angle of rotation is specified in radians by the variable A1 in line number 170. The variables X and Y represent the coordinate locations x and y, while the variables X1 and Y1 represent the coordinate locations x' and y'. The values of X and Y are plotted relative to the coordinates of the graphic template as line number 3020 is executed.

```
90  REM            *****  Program 2.1  *****
95  REM               Rotation of Coordinates
100 REM
         *****  set up graphics characteristics  *****

110 SCREEN 2 : CLS : XO = 320 : YO = 100 : SX = 1.5 :
SY = SX/2.25
150 REM
            *****  specify initial conditions  *****

160 R0 = 8
170 A1 = 1.57
300 REM    *****  set up screen display  *****

310 Y1 = 0 : REM draw horizontal axis
320 FOR X1 = -110 TO 110 STEP 2
330 XS = XO + SX*X1 : YS = YO - SY*Y1 : PSET (XS,YS)
340 NEXT X1
350 X1 = 0 : REM draw vertical axis
360 FOR Y1 = -100 TO 100 STEP 1.5
370 XS = XO + SX*X1 : YS = YO - SY*Y1 : PSET (XS,YS)
380 NEXT Y1
390 REM draw coordinate grid
400 FOR X1 = -100 TO 100 STEP 10
410 FOR Y1 = -90 TO 90 STEP 10
```

2.2 ROTATION OF COORDINATES

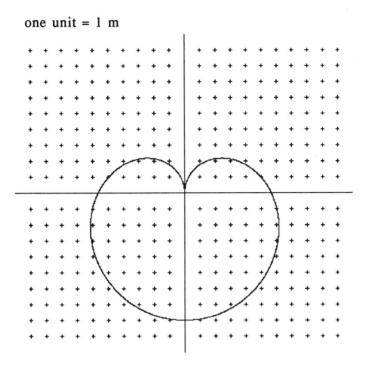

Figure 2.3 The orientation of the figure is affected by a rotation of coordinates.

```
420 XS = XO + SX*X1 : YS = YO - SY*Y1 : PSET (XS,YS)
430 NEXT Y1
440 NEXT X1
450 SC = 1 : REM scale for grid in meters
460 SX = 10*SX/SC : SY = SX/2.25
470 LOCATE 1,55 : PRINT "one unit =";SC;"m"
1000 REM
         *****  calculations and plotting  *****

1010 FOR TH = 0 TO 6.28 STEP .02
1020 R = R0*SIN(TH/2)
1030 X1 = R*COS(TH)
1040 Y1 = R*SIN(TH)
1050 GOSUB 3000
1060 NEXT TH
1100 END
2990 REM
         *****  transformation subroutine  *****

3000 X = X1*COS(A1) - Y1*SIN(A1)
3010 Y = X1*SIN(A1) + Y1*COS(A1)
3020 XS = XO + SX*X : YS = YO - SY*Y : PSET (XS,YS)
3030 RETURN
```

2.3 Translation of coordinates

The location of the origin of coordinates is changed while the orientation of the coordinate axes is preserved by a translation of coordinates. In Figure 2.4(a), the point P has coordinates (x,y) with respect to the coordinate axes of the figure. However, in Figure 2.4(b), the same point P has new coordinates (x',y') with respect to the coordinate axes because the origin of coordinates has undergone translation to a new location. Even though the two sets of axes have the same orientation, the respective origins of the coordinates do not share a common location.

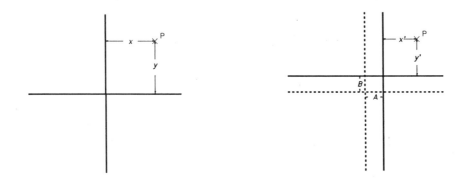

Figure 2.4(a)   Figure 2.4(b)

The coordinate axes of both systems are shown in Figure 2.5. Using this figure, transformation equations can be derived to relate (x,y) (the coordinates of P with respect to the axes x and y) to (x',y') (the coordinates of P with respect to the axes x' and y') in terms of A and B, which are the components of the translation of the origin of coordinates parallel to the x- and y-axes, respectively. (See Exercise 2.5.)

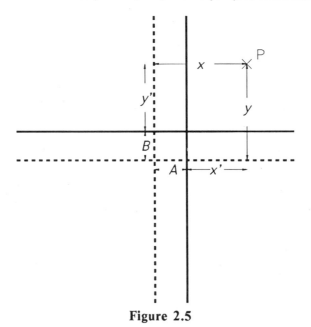

Figure 2.5

2.3 TRANSLATION OF COORDINATES

$$x = x' + A \qquad (2.3)$$

$$y = y' + B \qquad (2.4)$$

Using these transformation equations, a figure can be moved relative to the computer screen without altering the orientation or shape of the figure. As with the rotation of figures described in Section 2.2, the transformation equations are applied after the coordinates of each point of the figure are calculated. As the points corresponding to these new coordinate locations are plotted, the figure that was originally defined relative to a translated coordinate system in terms of points labelled (x',y') is now drawn relative to the graphic template of the computer screen in terms of points labelled (x,y).

This technique is applied in Program 2.2 to plot a version of the ellipse drawn by Program A.3 in Appendix A with the ellipse moved so that it is centered in the first quadrant of the coordinate system. Equations 2.3 and 2.4, the transformation equations, are applied in line number 3000 and in line number 3010, respectively. The components of the translation A and B (the variables XT and YT in the computer program) are specified in line numbers 170 and 180. The variables X and Y represent the coordinate locations x and y, while the variables X1 and Y1 represent the coordinate locations x' and y'. The values of X and Y are plotted relative to the coordinate system of the graphic template as line number 3020 is executed.

```
 90 REM             ***** Program 2.2 *****
 95 REM             Translation of Coordinates
100 REM
        *****  set up graphics characteristics  *****

110 SCREEN 2 : CLS : XO = 320 : YO = 100 : SX = 1.5 : SY = SX/2.25
150 REM
             *****  SPECIFY INITIAL CONDITIONS  *****
160 C = 8 : D = 4
170 XT = 3
180 YT = 3
300 REM
             *****  set up screen display  *****

310 Y1 = 0 : REM draw horizontal axis
320 FOR X1 = -110 TO 110 STEP 2
330 XS = XO + SX*X1 : YS = YO - SY*Y1 : PSET (XS,YS)
340 NEXT X1
350 X1 = 0 : REM draw vertical axis
360 FOR Y1 = -100 TO 100 STEP 1.5
370 XS = XO + SX*X1 : YS = YO - SY*Y1 : PSET (XS,YS)
380 NEXT Y1
390 REM draw coordinate grid
400 FOR X1 = -100 TO 100 STEP 10
```

12   2 · TRANSFORMATIONS OF COORDINATES

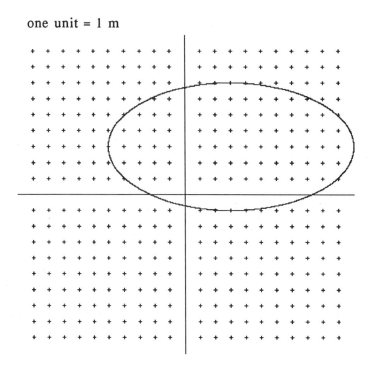

Figure 2.6 The position of the figure is determined by a translation of coordinates.

```
410 FOR Y1 = -90 TO 90 STEP 10
420 XS = XO + SX*X1 : YS = YO - SY*Y1 : PSET (XS,YS)
430 NEXT Y1
440 NEXT X1
450 SC = 1 : REM scale of screen grid in meters
460 SX = 10*SX/SC : SY = SX/2.25
470 LOCATE 1,55 : PRINT "one unit =";SC;"m"
1000 REM
             *****  calculations and plotting  *****

1010 FOR TH = 0 TO 6.28 STEP .01
1030 X1 = C*COS(TH)
1040 Y1 = D*SIN(TH)
1050 GOSUB 3000
1060 NEXT TH
1100 END
2990 REM
             *****  transformation subroutine  *****

3000 X = X1 + XT
3010 Y = Y1 + YT
3020 XS = XO + SX*X : YS = YO - SY*Y : PSET (XS,YS)
3030 RETURN
```

2.4 Three-Dimensional Transformations

By applying a simple set of transformation equations, functions of three dimensions (x',y',z') can be represented on the two-dimensional computer screen. The x'-axis and y'-axis are each drawn as in the figures created earlier in this chapter. An additional axis (the z'-axis) is represented by a line drawn at an angle ϕ with respect to the horizontal axis. As seen in Figure 2.7, a point $(0,0,z')$ on the $+z'$-axis is thus plotted a distance $z' \cos \phi$ in the $-x$ direction from the origin of coordinates and a distance $-z' \sin \phi$ below the origin of coordinates.

The locations of points (x',y',z'), which are not necessarily on the z'-axis, are determined by transformation equations.

$$x = x' - z' \cos \phi \tag{2.5}$$

$$y = y' - z' \sin \phi \tag{2.6}$$

These transformation equations are applied in Program 2.3 to create a graphic template for plotting in three dimensions. In this graphic template, the x'- and y'-axes are located in the same orientation relative to the computer screen as those of the two-dimensional graphic template created using Program A.1. The z'-axis is drawn at an angle ϕ with respect to the horizontal axis as described above. A coordinate grid is drawn in the $(x'-z')$ plane. The angle ϕ, represented by the variable PH, is specified in radians in line number 160. Equations 2.5 and 2.6 are applied in the subroutine beginning at line number 3000. As in Program A.1, the horizontal axis is drawn by execution of the for-next loop beginning at line number 320 and the vertical axis is drawn by execution of the for-next loop beginning at line number 360. The z-axis is then drawn by execution of the for-next loop beginning at line number 395.

```
 90 REM              *****  Program 2.3  *****
 95 REM          Three-Dimensional Transformations
100 REM
          *****  set up graphics characteristics  *****

110 SCREEN 2 : CLS : XO = 320 : YO = 100 : SX = 1.5 : SY = SX/2.25
150 REM
              *****  specify initial conditions  *****
160 PH = .5
300 REM
                 *****  set up screen display  *****
310 Y1 = 0 : Z1 = 0 : REM draw horizontal axis
320 FOR X1 = -110 TO 110 STEP 2
330 GOSUB 3000
340 NEXT X1
350 X1 = 0 : Z1 = 0 : REM draw vertical axis
360 FOR Y1 = 0 TO 110 STEP 1.5
370 GOSUB 3000
380 NEXT Y1
```

2 · TRANSFORMATIONS OF COORDINATES

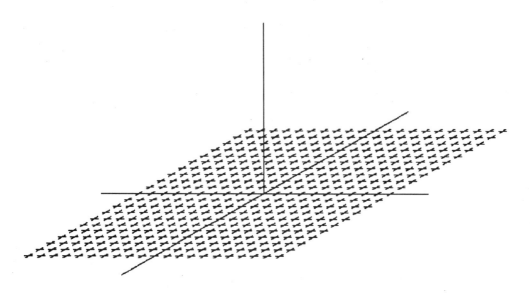

Figure 2.7 A three-dimensional coordinate system created with Program 2.3.

```
390  X1 = 0 : Y1 = 0
395  FOR Z1 = -110 TO 110 STEP 2
400  GOSUB 3000
405  NEXT Z1
410  Y1 = 0
415  FOR X1 = -100 TO 100 STEP 10
420  FOR Z1 = -100 TO 100 STEP 10
425  GOSUB 3000
430  NEXT Z1
435  NEXT X1
450  SC = 1 : REM scale of screen grid in meters
460  SX = 10*SX/SC : SY = SX/2.25
470  LOCATE 1,55 : PRINT "one unit =";SC;"m"
1000   REM        *****  calculations and plotting  *****

1100 END
2990 REM     *****  transformation subroutine  *****

3000 X = X1 - Z1*COS(PH)
3010 Y = Y1 - Z1*SIN(PH)
3020 XS = XO + SX*X : YS = YO - SY*Y : PSET (XS,YS)
3030 RETURN
```

2.4 THREE-DIMENSIONAL TRANSFORMATIONS

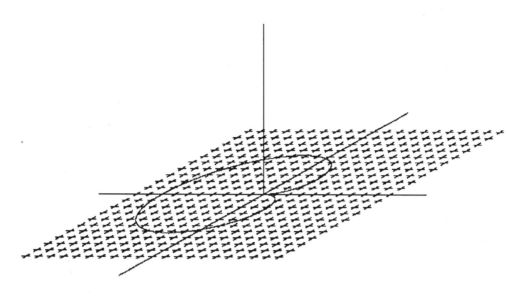

one unit = 1 m

Figure 2.8 The cardioid function is plotted in the $(x'\text{-}z')$ plane.

Mathematical functions can be plotted relative to this coordinate system. A set of values for a coordinate location (x',y',z') is first calculated. The location of this point relative to the computer screen is then determined by using the transformation equations (Eqs 2.5 and 2.6) by execution of the subroutine beginning at line number 3000. Other coordinates are determined and plotted in succession. To illustrate this procedure, the program lines below can be added to Program 2.3 to plot the cardioid function $(r = r_0 \sin \theta/2)$ of Figure A.5 in the $(x'\text{-}z')$ plane.

```
1010 Y1 = 0: R0 = 80
1020 For TH = 0 to 6.28 step .01
1030 R = R0*SIN(TH/2)
1040 X1 = R*COS(TH)
1050 Z1 = R*SIN(TH)
1060 GOSUB 3000
1070 NEXT TH
```

2.5 Galilean Transformations

The transformation equations for rotations and translations can only relate coordinate locations to coordinate systems that are at rest with respect to one another. Other transformation equations can be found to relate coordinate locations to coordinate systems in relative motion with respect to one another. For the special case of two inertial reference frames S and S' separating at a constant speed u, the Galilean transformation equations (also called the classical space-time transformation equations) relate the two reference frames. The transformation equations for the Galilean transformation are found in Eqs 2.4 - 2.7. In order for the equations to take this simple form, the coordinate axes of the two systems must be parallel and the origins of coordinates must coincide when $t = t' = 0$. The system S' is considered to be moving at velocity u in the positive x direction relative to S.

$$x = x' + ut \qquad (2.7)$$

$$y = y' \qquad (2.8)$$

$$z = z' \qquad (2.9)$$

$$t = t' \qquad (2.10)$$

These transformation equations can be used to describe the motion of objects with respect to inertial coordinate systems in relative motion. As with rotations and translations of coordinates, the transformation equations are applied after the coordinates of the location of the moving object are determined.

Program A.4 of Section A.7 determines the path of a projectile by calculating the coordinates of the location of the projectile as a function of time. To create Program 2.4, the program was further altered to apply the Galilean transformation equations to relate the path of the projectile to a different inertial reference frame with relative motion. The relative speed u of reference frames S and S' is specified in line number 190.

As in Program A.7, the position of the projectile is first determined relative to the S' reference frame in terms of the variables X1 and Y1. The horizontal position of the projectile relative to the reference frame S' is determined in line number 1020. The vertical position of the projectile is determined in line number 1030. By execution of the subroutine beginning at line number 3000, the Galilean transformation equations are applied to determine the position relative to the reference frame S in terms of the variables X and Y. The path of the projectile relative to reference frame S is represented on the computer screen using the graphic template.

```
 90 REM              *****  Program 2.4  *****
 95 REM Projectile Motion and Galilean Transformations
100 REM
            *****  set up graphics characteristics  *****

110 SCREEN 2 : CLS : XO = 320 : YO = 100 : SX = 1.5 : SY = SX/2.25
150 REM
            *****  specify initial conditions  *****

160 V0 = 40  : REM initial velocity of projectile
170 A = .9   : REM angle at which projectile is fired
180 G = -9.8 : REM acceleration due to gravity
```

2.5 GALILEAN TRANSFORMATIONS

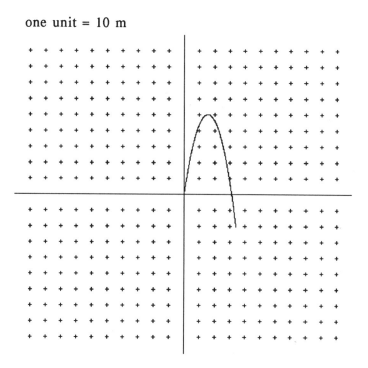

Figure 2.9 Projectile motion of Program 2.4 is shown relative to a moving observer for which $u = -20$ m/s.

```
190 U =-20: REM speed of (x1 , y1) coordinate system rela-
tive to computer
200 DT = .1
300 REM       *****  set up screen display  *****

310 Y1 = 0 : REM draw horizontal axis
320 FOR X1 = -110 TO 110 STEP 2
330 XS = XO + SX*X1 : YS = YO - SY*Y1 : PSET (XS,YS)
340 NEXT X1
350 X1 = 0 : REM draw vertical axis
360 FOR Y1 = -100 TO 100 STEP 1.5
370 XS = XO + SX*X1 : YS = YO - SY*Y1 : PSET (XS,YS)
380 NEXT Y1
390 REM draw coordinate grid
400 FOR X1 = -100 TO 100 STEP 10
410 FOR Y1 = -90 TO 90 STEP 10
420 XS = XO + SX*X1 : YS = YO - SY*Y1 : PSET (XS,YS)
430 NEXT Y1
440 NEXT X1
450 SC = 10
460 SX = 10*SX/SC : SY = SX/2.25
470 LOCATE 1,55 : PRINT "one unit =";SC;"m"
1000 REM      *****  calculations and plotting  *****
```

18 2 · TRANSFORMATIONS OF COORDINATES

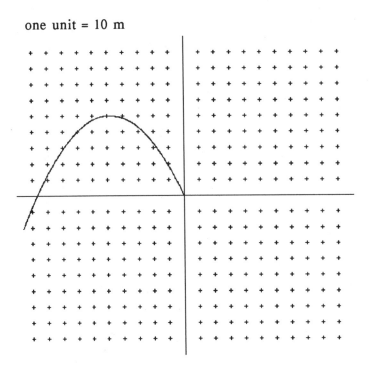

Figure 2.10 Projectile motion of Program 2.4 is shown relative to a moving observer for which $u = -40$ m/s.

```
1010 FOR T = 0 TO 7   STEP DT
1020 X1= V0*COS(A)*T
1030 Y1= V0*SIN(A)*T + .5*G*(T^2)
1040 GOSUB 3000
1050 NEXT T
1100 END
2990 REM
         *****  transformation subroutine  *****

3000 X = X1 + U*T
3010 Y = Y1
3020 XS = XO + SX*X : YS = YO - SY*Y : PSET (XS,YS)
3030 RETURN
```

Using the Galilean transformation equations, motion of an object relative to reference frame S′ is related to reference frame S (S′ is moving at speed u relative to S). As a further example, Program A.3 can be altered to display the path relative to reference frame S for an object moving in a circular path at a uniform angular velocity relative to the reference frame S′. The angular position of the object at any instant is $\theta = v_a t$. By changing the subroutine of Program A.3, the Galilean transformation can be applied to create a figure on the computer screen representing the path seen by an observer at rest relative to the S reference frame. The apparent path is shown in Figure 2.11.

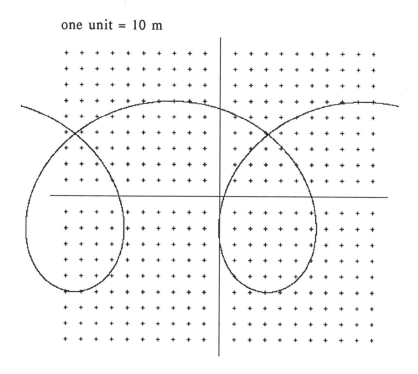

Figure 2.11 The apparent shape of a circular orbit observed from a moving coordinate system.

2.6 The Classical Doppler Effect

Sound waves travelling through a stationary medium can be represented using Eq A.8. Program A.5 graphs this equation to depict a travelling wave at any instant t. In this program both the transmitter that produces the wave and the receiver that detects the wave are assumed to be at rest relative to the medium through which the wave propagates.

The program can be modified to apply the Galilean transformation to depict these waves as detected by a moving receiver. For the special case of sound waves, the waves travel at a speed v_s relative to the medium through which the waves are propagated. In Program 2.5, this speed is specified by the variable C in line number 170 as 331 m/s, the speed of sound in air at standard temperature and pressure. Thus, the figure represents a sound wave travelling through air. The program applies the Galilean transformation to graph the wave as seen by a receiver moving through the air at a speed $-u$ parallel to the direction of propagation of the sound wave. In this transformation, the receiver is at rest relative to the reference frame S, and the air is at rest relative to reference frame S'. Thus, the sound receiver is moving relative to the air through which the sound wave is propagated, and the source of the sound is assumed to be stationary with respect to the air.

The speed of sound relative to the air is related to the wavelength and frequency of the sound waves observed at the transmitter by the relation $v_s = fL$. For the case of sound waves observed by a moving receiver, the wavelength L of the waves is the same for both the receiver and the transmitter. However, observers at the transmitter will see different values of speed and frequency than those observed at the receiver.

In Program 2.5, the variables X1 and T1 are associated with S', the reference frame at rest relative to the air through which the wave is propagated. The values specified for the speed C and the wavelength L are the values observed relative to this reference frame. Even though sound waves are in fact longitudinal waves, the wave is represented on the computer screen as a transverse wave. The amplitude A of the wave represents the pressure variations in the air through which the wave is propagated.

The time t and location x at which the wave is being detected by an observer at rest with respect to reference frame S are specified by execution of the nested for-next loops beginning at line number 1005 and line number 1010, respectively. Equation 2.7 is applied in line number 1020 to determine the location X1 at which the observation is made relative to the reference frame S'. The value of the wave displacement Y associated with X1 is calculated by execution of line number 1040. The value of the position X and the displacement of the wave Y are plotted relative to the graphic template of the computer screen in line number 1050.

The speed of the wave can be calculated from the figure on the computer screen. The distance s that the wave moves between successive drawings can be determined from the position of the figures relative to the graphic template. By dividing this distance by the time t between the drawings, the speed of the wave can be calculated using the equation $v = s/t$.

As seen in Figures 2.12 and 2.13, the speed of the wave is altered relative to the moving detector. If the speed of the detector through the air ($-u$) is positive so that the detector is moving away from the source, the measured speed of the wave is decreased. When the speed of the detector is negative, the detector is moving toward the source and the speed of the wave is increased. It is left as an exercise for the student to verify that the values of frequency and wave speed are in agreement with theoretical predictions. (See Exercise 2.11.)

```
 90 REM              ***** Program 2.5 *****
 95 REM              The Classical Doppler Effect
100 REM
         *****  set up graphics characteristics  *****

110 SCREEN 2 : CLS : XO = 320 : YO = 100 : SX = 1.5 : SY = SX/2.25
150 REM
              *****  specify initial conditions  *****

160 L = 6
170 C = 331
180 A1 = 5
190 U = 165
200 T1 = 0
210 T = 0
```

2.6 THE CLASSICAL DOPPLER EFFECT

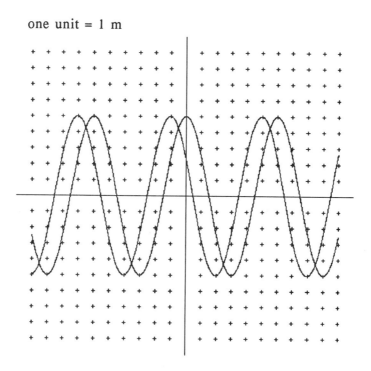

Figure 2.12 A sound wave with a speed of 331 m/s is detected by a receiver for which $u = 165$ m/s. The time between successive depictions of the wave is 0.01 s. The wave moves almost 5 m relative to the detector during this interval.

```
300 REM
                ***** set up screen display *****

310 Y1 = 0 : REM draw horizontal axis
320 FOR X1 = -110 TO 110 STEP 2
330 XS = XO + SX*X1 : YS = YO - SY*Y1 : PSET (XS,YS)
340 NEXT X1
350 X1 = 0 : REM draw vertical axis
360 FOR Y1 = -100 TO 100 STEP 1.5
370 XS = XO + SX*X1 : YS = YO - SY*Y1 : PSET (XS,YS)
380 NEXT Y1
390 REM draw coordinate grid
400 FOR X1 = -100 TO 100 STEP 10
410 FOR Y1 = -90 TO 90 STEP 10
420 XS = XO + SX*X1 : YS = YO - SY*Y1 : PSET (XS,YS)
430 NEXT Y1
440 NEXT X1
450 SC = 1 : REM scale for screen grid
460 SX = 10*SX/SC : SY = SX/2.25
470 LOCATE 1,55 : PRINT "one unit =";SC;"m"
```

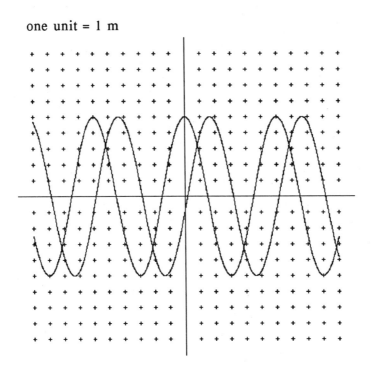

Figure 2.13 A sound wave with a speed of 331 m/s is detected by a receiver for which $u = -165$ m/s. The time between successive depictions of the wave is 0.01 s. The wave moves less than 2 m relative to the detector during this interval.

```
1000 REM      *****  calculate values and plot function  *****

1005 FOR T = 0 TO .02 STEP .01:REM note that wave moves at
the speed of sound
1010 FOR X = -10 TO 10 STEP .05
1020 X1 = X - U*T
1030 T1 = T
1040 Y = A1 * COS(6.28*(X1 - C*T1)/L)
1050 XS = XO + SX*X : YS = YO - SY*Y : PSET (XS,YS)
1060 NEXT X
1070 NEXT T
2000 END
```

Exercises

2.1 From the geometry of Figures 2.1(a) and 2.1(b), derive the transformation equations for rotations (Eqs 2.1 and 2.2).

2.2 Alter Program 2.1 to plot the polar functions of Exercise A.5 but rotated by 0.3 radians.

EXERCISES

2.3 Modify Program 2.1 to draw a family of ellipses with each successive ellipse rotated through an additional angle of 0.2 radians.

2.4 Alter Program 2.2 to plot the polar functions of Exercise A.5 with the center of the figure moved to the first quadrant of the graphic template.

2.5 a. Using Figure 2.4, derive Eq 2.3 and Eq 2.4.
b. Alter Program 2.2 to draw a family of four ellipses with an ellipse centered in each of the four quadrents of the graphic template.

2.6 a. Modify Program 2.2 to plot the cardioid function $r = r_0 \sin(\theta/2)$.
b. The set of transformation equations describing an inversion about the origin of coordinates is

$$x = -x' \; ; \; y = -y' \qquad (2.11)$$

Modify Program 2.2 to apply Eq 2.11 to plot an inverted figure of the cardioid function.
c. The set of transformation equations describing a mirror reflection about the y-axis is

$$x = x' \; ; \; y = -y' \qquad (2.12)$$

Modify Program 2.2 to apply Eq 2.12 to plot a mirror inversion of the cardioid function.

2.7 a. Derive transformation equations to produce an inversion about an arbitrary point (A,B).
b. Alter Program 2.2 to draw an ellipse inverted about some point other than the origin of the coordinate system.

2.8 a. Alter Program 2.3 to plot the cardioid function in the $(y\text{-}z)$ plane.
b. Alter Program 2.3 to plot the cardioid function in the $(x\text{-}y)$ plane.

2.9 a. Using Program 2.4, make measurements using the figure on the computer screen to verify that the horizontal component of velocity relative to the coordinate system of the screen is $v - u$.
b. Verify that the vertical component of velocity is unaltered by this coordinate transformation.

2.10 a. The moon orbits the earth in a nearly circular orbit with a period of 28 days. Alter Program A.3 to create a scale drawing of the orbit of the moon in the plane of the computer screen and centered on the screen.
b. Apply the Galilean transformation equations to illustrate the path of the moon in the same orientation as that created in (a) but as seen by an observer at rest with respect to the sun.

2.11 a. Using Program 2.5, make measurements using the figure to verify that the speed of the sound wave is $c + u$.

b. Adjust the time interval between successive drawings of the wave to determine the period of the wave by making the second figure superpose the first figure.

c. Measure the wavelength L using the figure on the computer screen. Using the value of the period determined in (b), calculate the frequency of the wave. From this, verify the relation $c = fL$.

2.12 Using Program 2.5, verify the relations for the observed frequencies and wavelengths for sound waves detected by a moving receiver.

Chapter 3

RELATIVISTIC TRANSFORMATIONS

3.1 Introduction

Einstein's two postulates provide the basis for the special theory of relativity.

1. The mathematical forms of the laws of physics are invariant in any inertial reference frame.
2. The speed of light is the same for any observer in any inertial reference frame.

Thus, the things about which all observers agree are the fundamental laws of nature and the speed of light. Contrary to everyday experience, however, observers don't necessarily agree on locations and dimensions in space and time. Distance measurements and time intervals are altered by relative motion. Worse yet, observers in different inertial reference frames can't agree on simultaneity of events. Events that occur simultaneously according to observers in one coordinate system may not appear to be simultaneous to observers in another coordinate system.

A set of transformation equations—called the Lorentz transformation equations—are used to relate observations made in different inertial coordinate systems. These equations are analogous to the Galilean transformation equations but are consistent with Einstein's postulates. The Lorentz transformation equations can be applied using the techniques of the previous chapter. When using these equations, particular care must be taken to properly prescribe the time at which observations are made in each coordinate system.

3.2 The Lorentz Transformation

The Lorentz transformation equations relate the coordinates of an event recorded relative to an inertial reference frame S' to those recorded relative to a second inertial reference frame S. An event occurs at a location designated (x,y,z) and at a time t when recorded by an observer who is at rest relative to reference frame S. This same event occurs at a location (x',y',z') and a time t' when recorded by an observer who is at rest with respect to reference frame S'.

Equations 3.1, 3.2, 3.3, and 3.4 are the Lorentz transformation equations that determine the coordinates (x',y',z') and time t' of an event relative to S' when the coordinates (x,y,z) and time t of the event are known relative to S. The relative speed u of the S' system is parallel to the x axis. In these equations, $\gamma = 1/(1 - u^2/c^2)^{1/2}$. For these equations to be valid, the origins O and O' of the reference frames are assumed to coincide when $t = t' = 0$.

$$x' = \gamma(x - ut) \qquad (3.1)$$

$$y' = y \qquad (3.2)$$

$$z' = z \qquad (3.3)$$

$$t' = \gamma(t - ux/c^2) \qquad (3.4)$$

The inverse transformations can be used to find the coordinates (x,y,z) and time t of an event relative to S when the coordinates (x',y',z') and time t' of the event are known relative to S'.

$$x = \gamma(x' + ut') \qquad (3.5)$$

$$y = y' \qquad (3.6)$$

$$z = z' \qquad (3.7)$$

$$t = \gamma(t' + ux'/c^2) \qquad (3.8)$$

In Program 3.1, the graphic template defined relative to the coordinate system S' is drawn on the computer screen as it would be recorded by observers at rest with respect to reference frame S. *The computer screen is thus considered to be stationary relative to S.* The locations of lighted pixels are determined by applying the Lorentz transformation at a time t to calculate the coordinates (x,y,z) corresponding to each point (x',y',z').

It is important to note that the figure drawn by this program does not represent the physical appearance of the graphic template recorded by a single observer at rest with respect to reference frame S, but instead represents a mapping of the grid that would be formed if simultaneous observations were made by observers at many different points of reference frame S. The physical appearance of objects recorded by a single observer is determined in Section 3.3.

In order to produce this figure, the locations of the points of the graphic template would be simultaneously recorded at a time t ($t = 0$ in this program) by a team of observers each of whom was associated with a different point in S. As a result, the time at which the observation is recorded is t at all points in S. Even if the observers associated with S (the computer screen) are careful to use sychronized clocks, observers in S' will not agree that all of the points were recorded simultaneously. The observers in S' will report that the time t' at which observations were made depended on the location of the observation in agreement with Eq 3.8. Thus, when calculations are made to find the coordinate locations observed relative to S, neither set of equations can be applied "as is." In the computer program, the coordinate locations (x',y',z') of the graphic template are known relative to S' but the time of observation t is known to be $t = 0$. Solving Eq 3.8, the time t' at which a coordinate is observed is thus

$$t' = t/\gamma - ux'/c^2 \qquad (3.9)$$

In Program 3.1, the graphic template is defined relative to S' in terms of variables X1, Y1, Z1, and T1. As the graphic template is drawn, the transformation equations are applied in the subroutine beginning at line number 3000. Equation 3.9 is used in line number 3000 to calculate the time t' (the variable T1 in the program). Using this value for t', Eq 3.5 is applied in line number 3010 to calculate x, the coordinate location relative to reference frame S (the reference frame of the computer screen). In the program, γ is represented by G. The location of each point of the graphic template is thus determined and plotted on the computer screen.

Both Figure 3.1 and Figure 3.2 were created using Program 3.1. The values chosen for u, the relative speed of the reference frames S and S', were $0.8c$ and $0.95c$. The Lorentz contraction of the graphic template is readily apparent in both figures. The figures can be used to verify the Lorentz contraction relations. (See Exercise 3.1.)

```
 90 REM              *****  Program 3.1  *****
 95 REM             The Lorentz Transformation
100 REM
     *****  set up graphics characteristics  *****

110 SCREEN 2 : CLS : XO = 320 : YO = 100 : SX = 1.5 : SY = SX/2.25
150 REM
                *****  initial conditions  *****

160 C = 3*10^8 : REM speed of light
170 U = .8*C : REM speed of (x1 , y1) coordinate system
relative to the computer
180 B = U/C
190 G = 1 / SQR(1 - B*B)
200 T = 0
```

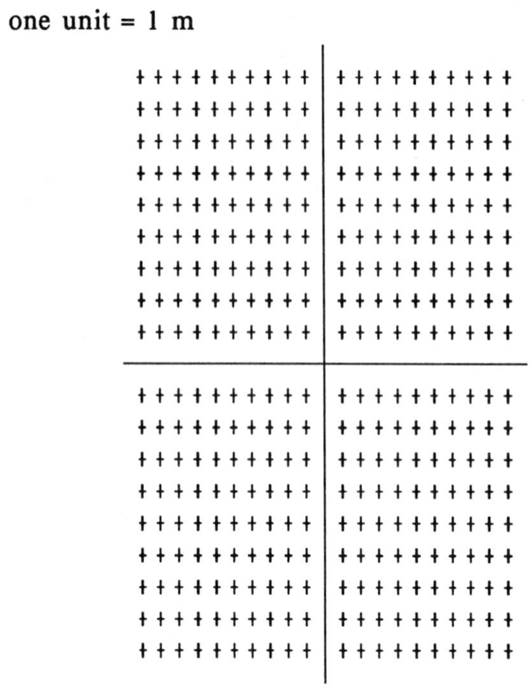

3 · RELATIVISTIC TRANSFORMATIONS

one unit = 1 m

```
+ + + + + + + + + + | + + + + + + + + + +
+ + + + + + + + + + | + + + + + + + + + +
+ + + + + + + + + + | + + + + + + + + + +
+ + + + + + + + + + | + + + + + + + + + +
+ + + + + + + + + + | + + + + + + + + + +
+ + + + + + + + + + | + + + + + + + + + +
+ + + + + + + + + + | + + + + + + + + + +
+ + + + + + + + + + | + + + + + + + + + +
+ + + + + + + + + + | + + + + + + + + + +
―――――――――――――――――――|―――――――――――――――――――
+ + + + + + + + + + | + + + + + + + + + +
+ + + + + + + + + + | + + + + + + + + + +
+ + + + + + + + + + | + + + + + + + + + +
+ + + + + + + + + + | + + + + + + + + + +
+ + + + + + + + + + | + + + + + + + + + +
+ + + + + + + + + + | + + + + + + + + + +
+ + + + + + + + + + | + + + + + + + + + +
+ + + + + + + + + + | + + + + + + + + + +
+ + + + + + + + + + | + + + + + + + + + +
```

Figure 3.1 A coordinate system moving at $0.8c$ illustrates Lorentz contraction.

```
300 REM
              *****   set up screen display    *****

310 Y1 = 0 : REM draw horizontal axis
320 FOR X1 = -110 TO 110 STEP 2
330 GOSUB 3000
340 NEXT X1
350 X1 = 0 : REM draw vertical axis
360 FOR Y1 = -100 TO 100 STEP 1.5
370 GOSUB 3000
380 NEXT Y1
390 REM draw coordinate grid
400 FOR X1 = -100 TO 100 STEP 10
410 FOR Y1 = -90 TO 90 STEP 10
420 GOSUB 3000
430 NEXT Y1
440 NEXT X1
450 SC = 1 : REM scale for screen grid in meters
460 SX = 10*SX/SC :SY = SX/2.25
470 LOCATE 1,55 : PRINT "one unit =";SC;"m"
500 END
2990 REM
              *****   transformation subroutine    *****
```

3.3 GEOMETRIC APPEARANCE AT RELATIVISTIC SPEEDS

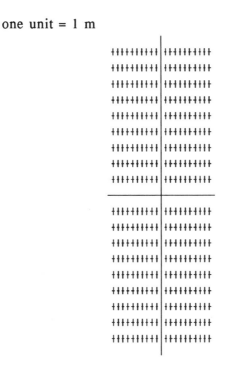

Figure 3.2 A coordinate system moving at $0.95c$ illustrates Lorentz contraction.

```
3000 T1 = (T/G) - ((U*X1) / (C*C))
3010 X = G*(X1 + U*T1)
3020 Y = Y1
3030 XS = XO + SX*X : YS = YO - SY*Y : PSET (XS,YS)
3040 RETURN
```

3.3 Geometric Appearance at Relativistic Speeds

The "mapping" procedure described in the previous section required the action of spatially separated observers who reported the results of simultaneous measurements made at time t relative to the reference frame S. These results were recorded by lighting pixels on the computer screen to represent the observations recorded by an observer associated with each pixel. This operation is quite different than the observation of an object by a single observer.

The computer program developed in this section produces projection drawings which represent the appearance of extended two-dimensional objects confined to the $(x'-y')$ plane of reference frame S' moving relative to a single observer at rest with respect to reference frame S. During the observation, a single observer is located on the $+z$-axis a distance d from the origin of reference frame S. The computer screen is considered to be the $(x-y)$ plane, so the observer is located in front of the computer screen at the point $(0,0,d)$. The reference frame S' moves in the $+x$ direction with a velocity u relative to S. The origins of coordinates O and O' coincide at the instant $t = t' = 0$. The

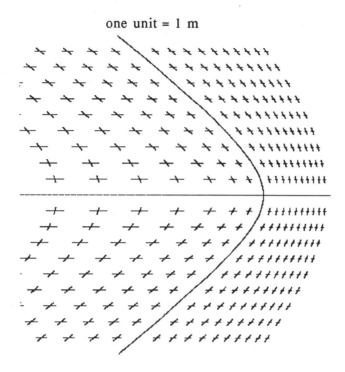

Figure 3.3 The geometrical appearance of a coordinate system moving at 0.8c was created using Program 3.2.

instant of observation occurs when the origins of the two systems momentarily coincide. As in photography, light reaching the observer from each point of the field of view is recorded at that moment.

For example, light emitted from O′, the origin of the reference frame S′, will appear to originate from O, the origin of the reference frame S, when $t = 0$ and will reach the observer when $t = d/c$ because the light travels the distance d from the origin to the observer at the speed of light c. Light from any other point $(x',y',0)$ will appear to the observer to have originated from a point $(x,y,0)$ relative to the reference frame S. The relation between the coordinates of the points $(x',y',0)$ and $(x,y,0)$ can be determined from the Lorentz transformation. The light travels a distance $(x^2 + y^2 + d^2)^{1/2}$ from the point $(x,y,0)$ before reaching the observer at the point $(0,0,d)$. Because of the greater length of its path before reaching the observer, the light must be emitted before the light from O′ in order to be observed at the same instant as the light from O′. If the light from the origin $(0,0,0)$ is to be received by the observer at $(0,0,d)$ when $t = d/c$, then the light from a point $(x,y,0)$ must be emitted when $t = d/c - (x^2 + y^2 + d^2)^{1/2}/c$ in order to be recorded simultaneously.

The Lorentz transformation is used to derive an equation to calculate the location $(x,y,0)$ from which the light from the point $(x',y',0)$ appeared to originate. For this calculation, the transformation equations are

$$x' = \gamma(x - ut); \quad y' = y; \quad z' = z$$

3.3 GEOMETRIC APPEARANCE AT RELATIVISTIC SPEEDS

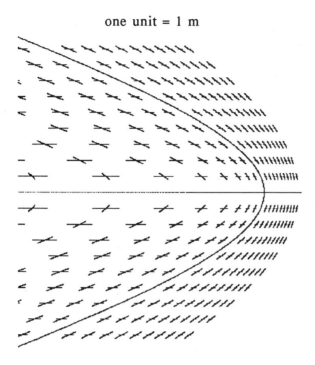

Figure 3.4 The geometrical appearance of a coordinate system moving at $0.95c$ was created using Program 3.2.

Substituting t, the time at which the light was emitted, into the set of equations yields a set of equations relating $(x',y',0)$, the location of an observed point relative to S', to $(x,y,0)$, the location in reference frame S from which the light appeared to originate.

$$x' = \gamma(x - u[d/c - (x^2 + y^2 + d^2)^{1/2}/c]) \tag{3.10}$$

$$y' = y \tag{3.11}$$

This set of equations can be solved to express x and y in terms of x' and y'. The coordinate $(x,y,0)$ is plotted to produce a drawing representing the observed position of the point $(x',y',0)$ recorded by a single observer at rest relative to reference frame S. (For more information about the derivation and application of these transformation equations, see the article by G. D. Scott and M. R. Viner, *American Journal of Physics*, July 1965.)

$$x = \gamma\{(x' + \gamma ud/c) - u[(x' + \gamma ud/c)^2 + y'^2 + d^2]^{1/2}/c\} \tag{3.12}$$

$$y = y' \tag{3.13}$$

In Program 3.2, Eqs 3.12 and 3.13 are used in the transformation subroutine to transform the graphic template defined relative to S' into a figure representing the observation of the graphic template made by a single observer at rest with respect to S and located at $(0,0,d)$ at the instant $t = t' = 0$. The computer determines the location of

one unit = 1 m

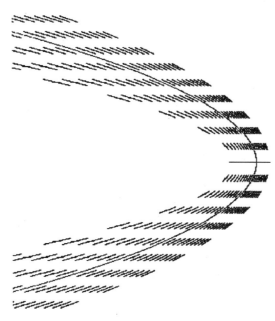

Figure 3.5 The geometrical appearance of a coordinate system moving at 0.995c was created using Program 3.2.

the points of the graphic template of S' as in earlier programs. The transformation equations are then applied in the subroutine at line number 3000 and the resulting coordinates are plotted.

```
 90 REM              *****  Program 3.2  *****
 95 REM     Geometrical Appearance of a Moving Object
100 REM
         *****  set up graphics characteristics  *****

110 SCREEN 2 : CLS : XO = 320 : YO = 100 : SX = 1.5 : SY = SX/2.25
150 REM         *****  initial conditions  *****

160 C = 3*10^8 : REM speed of light
170 U = .8*C: REM speed of (x1 , y1) coordinate system relative to the computer
180 B = U/C
190 G = 1 / SQR(1 - B*B)
200 D = 20
300 REM     *****  set up screen display  *****
310 Y1 = 0 : REM draw horizontal axis
320 FOR X1 = -110 TO 110 STEP 2
```

```
330 GOSUB 3000
340 NEXT X1
350 X1 = 0 : REM draw vertical axis
360 FOR Y1 = -100 TO 100 STEP 1.5
370 GOSUB 3000
380 NEXT Y1
390 REM draw coordinate grid
400 FOR X1 = -100 TO 100 STEP 10
410 FOR Y1 = -90 TO 90 STEP 10
420 GOSUB 3000
430 NEXT Y1
440 NEXT X1
450 SC = 1 : REM scale for screen grid
460 SX = 10*SX/SC : SY = SX/2.25
470 LOCATE 1,55 : PRINT "one unit =";SC;"m"
500 END
2990 REM
          *****  transformation subroutine  *****

3000 X = G*((X1 + G*B*D) - B*((X1 + G*B*D)^2 + Y1^2 + D^2)^.5)
3010 Y = Y1
3020 XS = XO + SX*X : YS = YO - SY*Y : PSET (XS,YS)
3040 RETURN
```

3.4 Relativistic Kinematics

Motion relative to a particular reference frame is simulated using Program A.4 to graph the kinematics equations which describe the motion as a function of time. The Galilean transformation equations are applied in Program 2.4 to relate motion to reference frames which are moving with respect to one another. The program illustrates the paths of moving objects as they appear to observers associated with any chosen reference frame. This application of the Galilean transformation equations is valid because of the classical principle of relativity.

In order to illustrate motion with respect to reference frames for which the relative velocities are a significant fraction of the speed of light, Program 2.4 must be modified to incorporate the Lorentz transformation equations. The Lorentz transformation is applied to map the path of the moving object relative to other coordinate systems. The motion of an object relative to reference frame S' is defined in terms of coordinate locations (x',y') and time t'. The Lorentz transformation equations (Eqs 3.5 - 3.8) can be applied to determine the motion of the object relative to the reference frame S in terms of coordinate locations (x,y) and time t.

In Program 3.3, motion is defined relative to the coordinate system S' in terms of the variables X1, Y1, and T1. For this program, the motion described is that of an object moving in a straight line parallel to the x'-axis at a speed of $-0.25c$ relative to S'. After the position of the object is determined at an instant T1 by execution of line number 1020, the Lorentz transformation equations are applied in the subroutine beginning at line number 3000 to calculate the position relative to S at the instant T in terms of the variables X and Y. The path of the object relative to S is then plotted on the computer screen. The graphic template is drawn relative to the coordinate system S, which is the coordinate system of the computer screen. As in Program 3.1, reference frame S' moves at a speed u relative to S. The origins O and O' of the coordinate systems coincide when $t = t' = 0$.

```
 90 REM              *****   Program 3.3   *****
 95 REM                 Relativistic Kinematics
100 REM
             *****  set up graphics characteristics  *****

110 SCREEN 2 : CLS : XO = 320 : YO = 100 : SX = 1.5 : SY = SX/2.25
150 REM
                    *****   initial conditions   *****

160 C = 3*10^8 : REM speed of light
170 U =.375*C:REM speed of (x1 , y1) coordinate system relative to the computer
180 B = U/C
190 G = 1 / SQR(1 - B*B)
200 VX = -.25*C :REM horizontal component of speed of moving object
300 REM
                  *****   set up screen display   *****

310 Y1 = 0 : REM draw horizontal axis
320 FOR X1 = -110 TO 110 STEP 2
330 XS = XO + SX*X1 : YS = YO - SY*Y1 : PSET (XS,YS)
340 NEXT X1
350 X1 = 0 : REM draw vertical axis
360 FOR Y1 = -100 TO 100 STEP 1.5
370 XS = XO + SX*X1 : YS = YO - SY*Y1 : PSET (XS,YS)
380 NEXT Y1
390 REM draw coordinate grid
400 FOR X1 = -100 TO 100 STEP 10
410 FOR Y1 = -90 TO 90 STEP 10
420 XS = XO + SX*X1 : YS = YO - SY*Y1 : PSET (XS,YS)
430 NEXT Y1
440 NEXT X1
450 SC = 10^8
460 SX = 10*SX/SC : SY = SX/2.25
470 LOCATE 1,55 : PRINT "one unit =";SC;"m"
1000 REM
                 *****  calculations and plotting  *****

1010 FOR T1 = 0 TO 10 STEP .1
```

3.4 RELATIVISTIC KINEMATICS

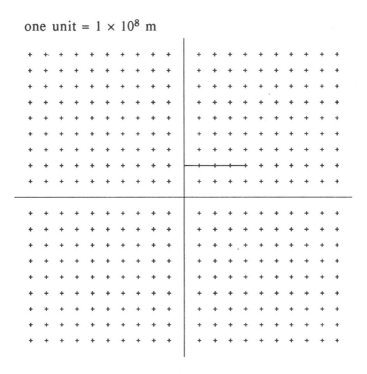

Figure 3.6 Motion in a straight line is illustrated using Program 3.3.

```
1020 X1 = VX*T1  : Y1 = 2*10^8 :REM motion of particle with
speed VX
1025 GOSUB 3000
1030 NEXT T1
1100 END
3000 REM
          *****   transformation subroutine   *****

3010 X = G*(X1 + U*T1)
3020 Y = Y1
3025 T = G*(T1 + (U*X1)/(C^2))
3030 XS = XO + SX*X : YS = YO - SY*Y : PSET (XS,YS)
3040 RETURN
```

Figure 3.6 can be used to verify the relativistic velocity transformation equations. The distance moved by the object can be measured directly from the figure. The elapsed time t was calculated in line number 3025. This value can be determined by the command PRINT T following execution of the program. Using this procedure, the particle is found to move a distance of 4×10^8 m in a time of 9.678 s. From these data, the velocity relative to S, the reference frame of the computer screen, is found to be $0.138c$. This result is consistent with the predictions of the relativistic velocity transformation equations but differs significantly from the value of $0.125c$ predicted by classical physics. (See Exercise 3.5.)

3.5 The Relativistic Doppler Effect

A light wave associated with reference frame S' is described by Eq 3.14. The wave travels with a speed c parallel to the x'-axis in the $+x'$ direction. The displacement of the wave y represents the strength of the electric field of the light wave at a point x' at an instant t'.

$$y = A \cos[2\pi(x' - ct')/L] \tag{3.14}$$

The properties of this function are explored in Section A.8. The classical Doppler effect for sound waves is investigated by applying the Galilean transformation in Program 2.5. In Program 3.4, the Lorentz transformation is applied to graph the displacement of the light wave relative to the reference frame S, the coordinate system of the computer screen. The reference frame S' moves at a speed u parallel to S. In order to plot the wave displacement relative to S, the Lorentz transformation is applied by using Eq 3.1 in line number 1020 to determine the coordinate x' of reference frame S' corresponding to the coordinate x of reference frame S. In addition, Eq 3.9 is applied in line number 1030 to calculate the time t' at which the calculation is made according to observers at rest with respect to reference frame S'. The value of y, the displacement of the wave corresponding to these values of x' and t' is calculated in line number 1040 using Eq 3.14. The displacement y and position x are then scaled and plotted on the computer screen to depict the shape of the wave observed relative to the reference frame S. The variables X1 and T1 of Program 3.4 represent x' and t', respectively.

In contrast to Figures 2.12 and 2.13 produced using the Galilean transformation equations, in Figure 3.7 the wavelength is seen to be altered when S and S' are in relative motion. This phenomenon is called the relativistic Doppler effect. However, unlike the sound waves depicted in Section 2.5, the speed of the wave is the same when determined relative to either reference frame S or reference frame S'. This can be verified by measuring the distance that the wave travels in reference frame S and dividing by the time t between successive calculations specified in line number 1005. No matter what the relative speed u between S and S', the speed of the light wave is found to be equal to $c (c = 3 \times 10^8$ m/s). This result is consistent with Einstein's first postulate. (See Exercise 3.6.)

```
 90 REM           *****  Program 3.4  *****
 95 REM           The Relativistic Doppler Effect
100 REM
           *****  set up graphics characteristics  *****

110 SCREEN 2 : CLS : XO = 320 : YO = 100 : SX = 1.5 : SY = SX/2.25
150 REM    *****  specify initial conditions  *****

160 C = 3*10^8
170 U = .5*C
180 B = U/C
```

3.5 THE RELATIVISTIC DOPPLER EFFECT   37

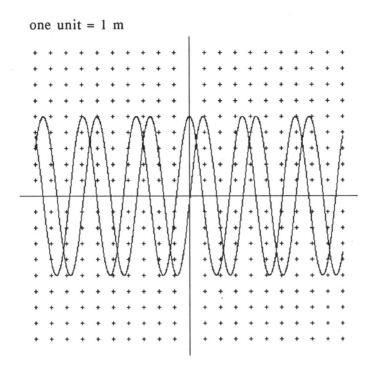

Figure 3.7 Because of the relativistic Doppler shift, the wavelength is contracted relative to reference frame S.

```
190 G = 1/ SQR(1 - B*B)
200 A1 = 5
210 L = 6
220 T1 = 0
230 T = 0
300 REM            *****  set up screen display  *****

310 Y1 = 0 : REM draw horizontal axis
320 FOR X1 = -110 TO 110 STEP 2
330 XS = XO + SX*X1 : YS = YO - SY*Y1 : PSET (XS,YS)
340 NEXT X1
350 X1 = 0 : REM draw vertical axis
360 FOR Y1 = -100 TO 100 STEP 1.5
370 XS = XO + SX*X1 : YS = YO - SY*Y1 : PSET (XS,YS)
380 NEXT Y1
390 REM draw coordinate grid
400 FOR X1 = -100 TO 100 STEP 10
410 FOR Y1 = -90 TO 90 STEP 10
420 XS = XO + SX*X1 : YS = YO - SY*Y1 : PSET (XS,YS)
430 NEXT Y1
440 NEXT X1
```

```
450 SC = 1 : REM scale for screen grid
460 SX = 10*SX/SC : SY = SX/2.25
470 LOCATE 1,55 : PRINT "one unit =";SC;"m"
1000 REM
           *****  calculate values and plot function  *****

1005 FOR T = 0 TO 6*10^-9 STEP 3*10^-9
1010 FOR X = -10 TO 10 STEP .05
1020 X1 = G*(X - U*T)
1030 T1 = T/G - U*X1/(C^2)
1040 Y = A1 * COS(6.28*(X1 - C*T1)/L)
1050 XS = XO + SX*X : YS = YO - SY*Y : PSET (XS,YS)
1060 NEXT X
1070 NEXT T
```

3.6 Relativistic Dynamics

The techniques used to study motion in the previous sections can be extended to study interactions between objects relative to different reference frames. Even though motion depends on the reference frame of the observer, the details of the motion must still be consistent with the underlying physical laws. For example, the law of conservation of momentum and the law of conservation of energy must be obeyed no matter what the reference frame of the observer.

The momentum and energy of an object with rest mass m_0 moving with speed v relative to an inertial reference frame are defined in Eqs 3.15 and 3.16.

$$p = m_0 v/(1-v^2/c^2)^{1/2} \tag{3.15}$$

$$E = m_0 c^2/(1-v^2/c^2)^{1/2} \tag{3.16}$$

Program 3.5 simulates a totally inelastic head-on collision between two identical objects each having a rest mass m'_0. The speed v of the colliding objects relative to their center of mass, the reference frame S', is specified in line number 200. The speed u of the reference frame S' relative to S, the reference frame of the computer, is specified in line number 170. The coordinates (x',y') of the moving objects at time t' before the collision are determined in line numbers 1020 and 1026. The collision occurs at the origin of coordinates when $t = t' = 0$. The coordinates (x',y') at the time t' after the collision are determined in line number 1050. The Lorentz transformation is applied in the subroutine beginning at line number 3000 to determine the coordinates (x,y) and time t relative to the reference frame of the observer. The coordinates (x,y) are scaled and plotted on the computer screen by execution of line number 3030.

The collision is shown in Figure 3.8 from the point of view of an observer at rest with respect to the center of mass of the two objects. Because identical objects are travelling with equal speeds but in opposite directions, in this reference frame the total momentum of the system is found to be zero both before and after the collision. In addition, all of the kinetic energy of the objects is converted to thermal energy as the objects remain at rest at the origin of coordinates after the collision.

3.6 RELATIVISTIC DYNAMICS

Before the collision depicted in Figure 3.8, the speed of each object is $0.5c$. Applying Eq 3.16, the energy of each object before the collision is found to be $1.15 m'_0 c^2$. The total energy of the system is therefore $2.30 m'_0 c^2$. The conservation of energy requires that the total energy of the system in this example must thus be $2.30 m'_0 c^2$ both before and after the collision. Because the speed v of the colliding objects after the collision is zero according to an observer at rest with respect to their center of mass, when Eq 3.16 is applied, the energy of the system is found to be $m_0 c^2$ where m_0 is the rest mass of the system after the collision. Equating this result with the energy of the system yields:

$$m_0 c^2 = 2.30 m'_0 c^2 \qquad (3.17)$$

This illustrates the equivalence of mass and energy as the rest mass of the system increased from $2.00 m'_0$ before the collision to be $2.30 m'_0$ after the collision. Before the collision, the total rest mass of the objects was $2.00 m'_0$. After the collision, the total rest mass of the system is found to be $2.30 m'_0$. The kinetic energy of the colliding objects before the collision was converted to rest mass in the stationary system after the collision. (See Exercise 3.8.)

By applying the Lorentz transformation in Program 3.5, it is possible to simulate this collision relative to other inertial reference frames. In Figure 3.9, the paths of the colliding particles are depicted as they appear when the collision occurs in a reference frame for which the center of mass is moving at a speed of $0.6c$ relative to the reference frame of the observer. It is left as an exercise for the student to apply Eqs 3.15 and 3.16 to verify that both the momentum and energy of the system are conserved and to calculate the energy-mass conversion when the collision is observed from reference frames other than the center of mass of the system. (See Exercise 3.8.)

As described in Section 3.4, the distance that each object moves can be measured directly from the figure on the computer screen. The value of t is calculated in line number 3025. The time from the beginning of the motion until the collision occurs at $t = 0$ must be determined separately for each of the colliding objects. These values are used to calculate the initial values of velocity and thus determine the initial momentum and energy of the system. Using the final value of t, the velocity after the collision can be similarly calculated.

```
 90 REM             ***** Program 3.5 *****
 95 REM                Relativistic Dynamics
100 REM
         *****  set up graphics characteristics  *****

110 SCREEN 2 : CLS : XO = 320 : YO = 100 : SX = 1.5 : SY = SX/2.25
150 REM         *****  initial conditions  *****

160 C = 3*10^8 : REM speed of light
170 U = .6*C
```

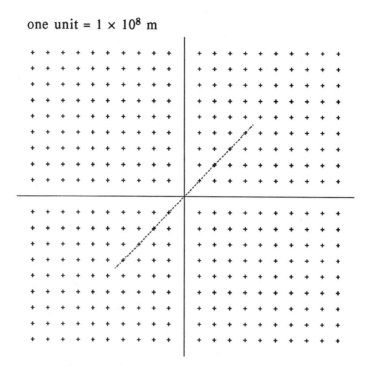

Figure 3.8 A totally inelastic head-on collision relative to an observer at rest with respect to the center of mass is depicted.

```
180 B = U/C
190 G = 1 / SQR(1 - B*B)
200 VX = .25*C : VY = .25* C
300 REM              *****  set up screen display  *****

310 Y1 = 0 : REM draw horizontal axis
320 FOR X1 = -110 TO 110 STEP 2
330 XS = XO + SX*X1 : YS = YO - SY*Y1 : PSET (XS,YS)
340 NEXT X1
350 X1 = 0 : REM draw vertical axis
360 FOR Y1 = -100 TO 100 STEP 1.5
370 XS = XO + SX*X1 : YS = YO - SY*Y1 : PSET (XS,YS)
380 NEXT Y1
390 REM draw coordinate grid
400 FOR X1 = -100 TO 100 STEP 10
410 FOR Y1 = -90 TO 90 STEP 10
420 XS = XO + SX*X1 : YS = YO - SY*Y1 : PSET (XS,YS)
430 NEXT Y1
440 NEXT X1
450 SC = 10^8 : REM scale for screen grid in meters
460 SX = 10*SX/SC : SY = SX/2.25
470 LOCATE 1,55 : PRINT "one unit =";SC;"m"
```

3.6 RELATIVISTIC DYNAMICS 41

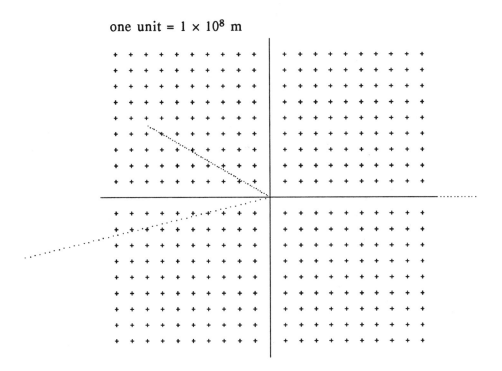

Figure 3.9 A totally inelastic head-on collision relative to an observer moving at 0.6c with respect to the center of mass is depicted.

```
1000 REM      ***** calculations and plotting    *****
1010 FOR T1 = -6   TO 0 STEP .1
1020 X1 = -VX*T1  : Y1 = -VY*T1 :REM particle moving at 5
units per second
1025 GOSUB 3000
1026 X1 = VX*T1 : Y1 = VY*T1
1027 GOSUB 3000
1030 NEXT T1
1040 FOR T1 = 0 TO 6   STEP .1
1050 X1 = 0 : Y1 =   0
1060 GOSUB 3000
1070 NEXT T1
1100 END
3000 REM
        *****   transformation subroutine    *****

3010 X = G*(X1 + U*T1)
3020 Y = Y1
3025 T = G*(T1 + (U*X1)/(C^2))
3030 XS = XO + SX*X : YS = YO - SY*Y : PSET (XS,YS)
3040 RETURN
```

Exercises

3.1 Using Program 3.1, make measurements using the figure on the computer screen to verify the Lorentz-Fitzgerald length contraction.

3.2 a. Alter Program 3.1 to plot an ellipse relative to the moving coordinate system with the major axis of the ellipse parallel to the horizontal axis.
b. Using the Lorentz-Fitzgerald length contraction equation, calculate the speed at which the ellipse will be mapped as a circle.
c. Verify this calculation by using the speed calculated in (b) as the speed of the coordinate system moving relative to the computer screen.

3.3 a. Alter Program 3.2 to illustrate the appearance of a circle centered at the origin of the coordinate system drawn relative to the moving reference frame.
b. Modify the program to illustrate the appearance of a circle centered on the x-axis at a point for which $x < 0$.
c. Modify the program to illustrate the appearance of a circle centered on the x-axis at a point for which $x > 0$. The appearance of objects moving at relativistic speeds changes as the objects move across the field of view of the observer.

3.4 Alter Program 3.2 to determine the appearance of the cardioid function plotted in the previous chapter.

3.5 a. Using Program 3.3, make measurements from the computer screen to verify the velocity transformation equations for velocity components parallel to the x-axis.
b. Modify Program 3.3 to verify the velocity transformation equations for velocity components perpendicular to the x-axis.

3.6 Apply Program 3.4 to verify that the speed of light is unaltered. To do this, draw the wave at two times, measure the distance travelled by the wave from the drawing on the computer screen. Divide the distance travelled by the elapsed time to determine the velocity of the wave relative to the coordinate system of the screen.

3.7 Using the figure on the computer screen, verify the relativistic frequency transformations. Determine the frequency by measuring the time that it takes for the wave to move a distance equal to one wavelength by drawing the wave at successive times.

3.8 a. Using Program 3.5, create Figure 3.9. Making measurements from the computer screen, measure the horizontal and vertical components of velocity before and after the collision. (See Exercise 3.5.)
b. Calculate the energy of the system before the collision.
c. Calculate the equivalent mass of the system after the collision assuming that energy was conserved by the energy-mass conversion in the collision.
d. Using Eq 3.15, verify that momentum is conserved in the collision.

Chapter 4

MATTER AND MOTION

4.1 Introduction

The development of the fundamental concepts and ideas of modern physics was stimulated by the results of historic experiments performed during the last half of the nineteenth century and the first half of the twentieth century. In order to interpret the results and to understand the significance of these experiments, it was necessary for physicists to visualize motion with speeds and dimensions outside the range of ordinary experience. Instead of classical experiments with rolling balls or falling objects, experimental physicists of this pioneering era studied the motion and properties of such subatomic exotica as canal rays, alpha rays, and cathode rays. The properties and nature of these particles beams—now recognized as ions, helium nuclei, and electrons, respectively—present intriguing puzzles even today.

In this chapter practical numerical methods are developed and applied in a study of matter and motion. Numerical methods are used to evaluate the ordinary differential equations that occur in the development of the ideas of modern physics. These powerful mathematical tools yield computer programs which both provide examples of the applications of numerical methods and simulate experimental results. With the aid of a computer, numerical "experiments" are devised so that the properties and motion of subatomic matter can be investigated. Using these computer programs, numerical versions of historic experiments can be performed.

As with the classical experiments featuring rolling balls and falling objects, Newton's laws of motion provide a starting point for interpreting the historic experiments of modern physics. For this purpose, matter—even on the microscopic scale—is considered to consist of particles whose motion is strictly in accordance with the laws of classical mechanics. In the chapter following this one, these same numerical methods are extended to explore the predictions of the quantum theory based on the wave-like properties exhibited by matter on the microscopic scale.

Even though the laws of classical mechanics were later shown to be inadequate for a full description of matter and motion on the microscopic scale, many important properties are revealed when Newton's laws are applied. Most importantly, matter is found to be quantized. Thus, matter—even with its incredible variety of forms—consists of structures formed from a few fundamental building blocks.

Newton's second law is expressed most concisely as a second-order differential equation.

$$F(r,v,t) = m d^2 r/dt^2 \qquad (4.1)$$

This equation describes the effect of a force F on the motion of a particle of mass m. To determine the motion of the particle it is necessary to solve this differential equation for the position vector r as a function of time. Because the applied force F is a vector quantity and can be almost any function of position, velocity, and time, no single general analytical method exists to provide solutions for this and the other differential equations encountered in the study of classical physics or modern physics. Each time a new type of differential equation is encountered, new mathematical techniques must be mastered to provide solutions. Even when analytical solutions are found, the mathematical form of the solutions often clouds the interpretation of the results.

Because new methods must be developed for each type of differential equation, problems must be restricted to easily solvable forms. This limits the useful examples of physical principles to the few cases yielding simple equations which can be solved with elementary techniques. Using numerical methods and ordinary microcomputers, a much larger number of examples—constrained only by the limits of our imagination—can be studied. This approach encourages concrete understanding of otherwise abstract and difficult ideas, especially when combined with graphical representations which simplify the interpretation of results.

4.2 Euler's Method

Euler's method is an elementary numerical technique which generates numerical solutions for any ordinary differential equation or any set of coupled ordinary differential equations. Although Euler's method can be applied to solve any ordinary differential equation, Newton's second law is used here to provide a simple and familiar example. After practice with this important example, extensions of Euler's method can easily be developed when new situations are encountered.

To describe motion in one dimension, Newton's second law reduces to a single ordinary second-order differential equation:

$$d^2 x/dt^2 = F(x,v,t)/m \qquad (4.2)$$

Applying the definition of a second derivative, Eq 4.2 becomes

$$\lim dt \to 0 \; [dx/dt_{t+dt} - dx/dt_t]/dt = F(x,v,t)/m \qquad (4.3)$$

4.2 EULER'S METHOD

In this equation, dx/dt_{t+dt} represents the velocity evaluated at the time $t + dt$ and dx/dt_t represents the velocity evaluated at the time t. Equation 4.3 is exact only in the case $dt \rightarrow 0$. However, the left side of the equation is a first-order approximation of d^2x/dt^2 for small finite values of dt. In principle, the accuracy of the approximation can be increased to any desired level by making the value of dt smaller.

For finite values of dt, Eq 4.3 becomes

$$dx/dt_{t+dt} = dx/dt_t + [F(x,v,t)/m]dt \qquad (4.4)$$

With Equation 4.4, dx/dt_{t+dt}, the velocity at the time $t+dt$ can be determined from knowledge of dx/dt_t, the velocity at an earlier time t.

Although the equation is written in a more abstract form, from an intuitive point of view Eq 4.4 amounts to the elementary result for motion with constant acceleration

$$v = v_0 + at \qquad (4.5)$$

The time interval t of Eq 4.5 is considered to be a short finite time increment dt in Eq 4.4. It is important to note that for the duration of the time increment dt, the acceleration is considered to be equal to the acceleration at the beginning of the increment: $F(x,v,t)/m_t$. During this short time increment, the acceleration is considered to be constant so that Eq 4.4 and Eq 4.5 are equivalent. The degree to which this assumption is true largely determines the accuracy of the calculated value of dx/dt_{t+dt}, the velocity at the end of the time increment dt.

The position at the instant $t + dt$ can be similarly determined from knowledge of the position at the instant t. The definition of the derivative provides a relation between velocity and position.

$$lim\ dt \rightarrow 0\ (x_{t+dt} - x_t)/dt = dx/dt \qquad (4.6)$$

For finite values of dt, Eq 4.6 becomes:

$$x_{t+dt} = x_t + (dx/dt_t)dt \qquad (4.7)$$

Equation 4.7 is equivalent to the elementary relation between position and time for motion with a constant velocity.

$$x = x_0 + vt \qquad (4.8)$$

The time interval t found in elementary equations is again replaced by the time increment dt in Eq 4.8. It is important to note that the velocity during the time increment is considered to be equal to the velocity at the beginning of the increment dx/dt_t. Although Eq 4.7 is exact only when the velocity is constant, in principle, the position at the instant $t + dt$ can be determined to any finite degree of accuracy if the time increment dt is made small enough.

If initial values of velocity and position are known, Eq 4.4 and Eq 4.7 can be used to determine the velocity and position a short time dt later. With knowledge of the calculated values of position and velocity at time $t + dt$, Eq 4.4 and Eq 4.7 can again be evaluated to find the velocity and position at the time $t + 2dt$.

$$dx/dt_{t+2dt} = dx/dt_{t+dt} + [F(x,v,t+dt)/m]dt \qquad (4.9)$$

$$x_{t+2dt} = x_{t+dt} + (dx/dt_{t+dt})dt \qquad (4.10)$$

Repeated application of this procedure yields values of position and velocity for subsequent times; $t + 3dt$, $t + 4dt$, $\cdots$. The number of repetitions of the calculation is determined by the magnitude of the time interval over which the motion is to be evaluated and by the size of the time increment dt.

4.3 Last Point Approximation

Conceptually, Euler's method is especially simple. In addition, this method is easy to apply in the context of computer programs. Many other methods (and families of methods) are available to provide numerical solutions to differential equations. Most other methods are more difficult to apply but often produce more accurate results or increase the efficiency of calculation. Other methods are described in texts on numerical methods such as those listed in the introductory chapter of this text.

The last-point approximation, a simple variant of Euler's method, often provides a practical way to improve the accuracy of calculations. Equations 4.11 and 4.12 are used to perform the successive calculations. Note that Eq 4.12 is equivalent to Eq 4.10 except that the speed at the end of the time increment dx/dt_{t+dt} (rather than the speed at the beginning of the increment dx/dt_t) is used to calculate the new position at the end of the increment x_{t+dt}. This change is the only distinction between Euler's method and the last-point approximation.

$$dx/dt_{t+dt} = dx/dt_t + [F(x,v,t)/m]dt \qquad (4.11)$$

$$x_{t+dt} = x_t + (dx/dt_{t+dt})dt \qquad (4.12)$$

When either Euler's method or the last-point approximation is used, the finite value of dt limits the accuracy of the results. This approximation—which is the fundamental step in the development of both Euler's method and the last-point approximation—necessarily restricts the accuracy of results. Because of this ultimate limit on accuracy, each of the repetitive calculations produces a small error that can accumulate as the calculations proceed. Error produced because of the finite values of the increments dt is called truncation error. In principle, truncation error can always be reduced by using smaller values of dt in calculations.

In practice, smaller values of dt may in fact reduce the accuracy of results and will certainly increase the computer time required for calculations. Each calculation made by the computer uses a limited number of significant figures so that each calculation has limited accuracy. Error of this type is called roundoff error. As the value of dt is decreased, the time interval t is broken into more increments. More calculations are performed and the level of roundoff error increases. Double precision calculations reduce roundoff error but are not necessary for reasonably accurate results with the programs of this text.

In many cases accumulated truncation errors are bounded when the last-point approximation is used. (See A. Cromer, *American Journal of Physics*, May 1981.) The following computer programs are all based on the last-point approximation. Clarity, accuracy, and speed were all considered in the development of the programs. However, if highly accurate results are required, more elaborate methods such as the fourth-order Runge-Kutta method for evaluating ordinary differential equations should be considered.

Both Euler's method and the last-point approximation can be applied when the differential equations to be evaluated contain variables other than position and time. The dependent variable of the differential equation is substituted for position x and the independent variable replaces the time t. Either method can then be applied as described.

4.4 Motion of an Oil Droplet

In 1909, Robert A. Millikan successfully completed a Nobel prize winning experiment to determine e, the value of the charge on the electron. The experiment also established that electric charge is quantized and permitted the determination of the mass of the electron. To perform this feat, Millikan studied the motion of charged oil droplets falling through air. The motion of the oil droplets was altered by electric fields applied parallel to the paths of the falling droplets. Analysis of the motion enabled Millikan to calculate the electric charge of the oil droplets.

The forces on an oil droplet include the weight of the droplet and the upward buoyant force of the air. The downward force acting on the droplet is

$$F_g = mg \qquad (4.13)$$

where m is the mass of the oil droplet and g is the acceleration due to gravity ($g = -9.8$ m/s^2).

In accordance with the principles of fluid statics, the upward buoyant force of the air is equal to the weight of the air displaced by the oil droplet. The weight of the air displaced by the oil droplet is equal to m_a, the mass of the air occupying a volume equal to the volume of the oil droplet, multiplied by g.

$$F_b = -m_a g \qquad (4.14)$$

In addition to its weight and the buoyant force of the air, the motion of a falling oil droplet is also affected by the force of fluid friction F_v which is proportional to the velocity of the oil droplet. The magnitude of the force of fluid friction is determined from Stoke's law.

$$F_v = -bv \qquad (4.15)$$

In this equation, the negative sign indicates that the direction of the friction force is opposite that of the velocity. The coefficient b is proportional to the product of the radius of the spherical oil droplet and the viscosity coefficient of the air.

All of the forces affecting an oil droplet falling vertically under the influence of gravity act parallel to the vertical axis. The net force affecting the falling oil droplet is thus equal to the sum of these forces.

$$F = (m - m_a)g - bv \tag{4.16}$$

Applying Eq 4.1, Newton's second law, the equation of motion of the falling droplet becomes

$$(m - m_a)g - bv = md^2y/dt^2 \tag{4.17}$$

When written in the form of Eq 4.2, this equation becomes

$$d^2y/dt^2 = [(m - m_a)g - bv]/m \tag{4.18}$$

Applying the last-point approximation in the form of Eq 4.11 and Eq 4.12 yields

$$dy/dt_{t+dt} = dy/dt_t + \{[(m - m_a)g - bv]/m\}dt \tag{4.19}$$

$$y_{t+dt} = y_t + (dy/dt_{t+dt})dt \tag{4.20}$$

Equations 4.19 and 4.20 are used in Program 4.1 to determine the motion of the falling oil droplet. Typical values for the mass of the oil droplet m and the mass of the air displaced by the oil droplet m_a are specified by the variables M and MA in line number 170 and line number 180, respectively. The velocity at the beginning of the time increment dy/dt_t is represented by the variable V, while the velocity at the end of the time increment dy/dt_{t+dt} is represented by the variable V1. Similarly, the position of the falling droplet at the beginning of the time increment y_t is represented by the variable Y and the position of the droplet at the end of the time increment y_{t+dt} is represented by the variable Y1. Equations 4.19 and 4.20, which are derived from the equation of motion of the falling droplet, are evaluated in line number 1020 and line number 1030, respectively. The results of the calculations are printed on the computer screen as line number 1040 is executed. After the results are displayed, the values of the variables V and X are redefined in line number 1030 in order to use the newly determined values V1 and X1 for the values of velocity and position at the beginning of the subsequent time increment. The for-next loop then advances the value of the variable t by a chosen increment dt. The calculation is repeated over the time interval specified in line number 1010.

```
 90 REM        *****  Program 4.1  *****
 95 REM        Motion of an Oil Droplet
150 REM
     *****  specify initial conditions   *****

155 G = -9.8
160 V = 0
170 M = 5*10^-15
180 MA = 7*10^-18
200 B = 3.8*10^-10
210 DT = .000001
220 X = 2*10^-9 : Y = 0
```

```
1000 REM
         *****  calculations and printing  *****

1010 FOR T = 0 TO .0001 STEP DT
1020 V1 = V +((((M - MA)*G)-(B*V))/M)*DT
1030 Y1 = Y + V1*DT
1040 PRINT V,Y,T
1050 Y = Y1 : V = V1
1060 NEXT T
1100 END
```

Program 4.1 generates tables of numbers which represent the position and velocity of the falling oil droplet. After a time the velocity of the falling oil droplet reaches a constant value—the terminal velocity of the falling droplet. In order to simulate the appearance of the motion, the tabulated values are plotted on the computer screen relative to the graphics template (Figure 4.1). In Program 4.2, the position of the falling droplet is plotted as a function of time. Instead of printing the values of speed and position, the location of the falling droplet is scaled to the dimensions of the graphics template and plotted as the subroutine beginning at line number 3000 is executed.

```
 90 REM              *****  Program 4.2  *****
 95 REM         Motion of an Oil Droplet (Graphic)
100 REM
         *****  set up graphics characteristics  *****

110 SCREEN 2 : CLS : XO = 320 : YO = 100 : SX = 1.5 : SY = SX/2.25
150 REM  *****  specify initial conditions  *****

160 G = -9.8
170 M = 5*10^-15
180 MA = 7*10^-18
200 B = 3.8*10^-10
210 DT = .000001
220 X = 2*10^-9 : Y = 0
230 V = 0
300 REM
            *****  set up screen display  *****

310 Y1 = 0 : REM draw horizontal axis
320 FOR X1 = -110 TO 110 STEP 2
330 XS = XO + SX*X1 : YS = YO - SY*Y1 : PSET (XS,YS)
340 NEXT X1
350 X1 = 0 : REM draw vertical axis
360 FOR Y1 = -100 TO 100 STEP 1.5
```

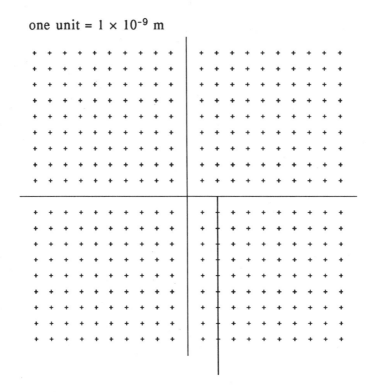

Figure 4.1 The motion of a falling oil droplet is depicted in the figure created by Program 4.2.

```
370 XS = XO + SX*X1 : YS = YO - SY*Y1 : PSET (XS,YS)
380 NEXT Y1
390 REM draw coordinate grid
400 FOR X1 = -100 TO 100 STEP 10
410 FOR Y1 = -90 TO 90 STEP 10
420 XS = XO + SX*X1 : YS = YO - SY*Y1 : PSET (XS,YS)
430 NEXT Y1
440 NEXT X1
450 SC = 10^-9: REM scale for screen grid in meters
460 SX = 10*SX/SC : SY = SX/2.25
470 LOCATE 1,55 : PRINT "one unit =";SC;"m"
1000 REM
            *****   calculations and plotting   *****

1010 FOR T = 0 TO .0001 STEP DT
1020 V1 = V +((((M - MA)*G)-(B*V))/M)*DT
1030 Y1 = Y + V1*DT
1040 GOSUB 3000
1050 Y = Y1 : V = V1
1060 NEXT T
1100 END
2990 REM
            *****   plotting subroutine   *****
```

```
3000 XS = XO +SX*X : YS = YO - SY*Y : PSET (XS,YS)
3010 RETURN
```

4.5 Radioactive Decay

Henri Becquerel was the first to detect the spontaneous emission of radiation by radioactive nuclei. In 1896, Becquerel noticed that radioactive materials were able to cause exposure of film which had not been reached by visible light. Early investigators felt that radioactive materials could continue to emit radiation indefinitely because the activity did not appear to diminish over a period of time. In a paper published in 1903, Ernest Rutherford and Fredrick Soddy reported, however, that the activity level of radioactive gas (radon-220) separated from thorium decreased by a factor of 2 every 52 seconds. This observation lead to the idea that radioactive decay is a statistical process.

According to this theory, each atom in a sample of radioactive material has the same probability of decay in a given time interval dt. This probability is equal to the product Adt where the variable A is called the activity constant of the radioactive material. The activity constant is determined empirically and depends only on the type of radioactive material being observed. The activity constant is thus different for each type of radioactive isotope.

As a sample of N atoms undergoes radioactive decay, the probability of decay for each atom in the sample is equal to Adt during the time interval dt. The total probability of decay among the atoms of the sample during a short time interval dt is increased by a factor N. The number of atoms dN by which the sample size is reduced by decay in time dt is thus $-NAdt$. This result is, of course, strictly correct only in the case that the time interval dt approaches zero.

$$dN = -NAdt \qquad (4.21)$$

Thus, the rate of decay of a sample of radioactive material is described by a first-order differential equation.

$$dN/dt = -NA \qquad (4.22)$$

Applying Eq 4.10, this equation becomes:

$$N_{t+dt} = N_t - NAdt \qquad (4.23)$$

The process described by Eq 4.23 is an elementary method of numerical integration. Equation 4.23 is applied in Program 4.3 to determine the number of atoms in a sample of radioactive material as a function of time. The number of atoms of the initial sample is specified in line number 160. The activity constant A is specified in line number 170. A for-next loop is applied to advance the time in increments of dt over the range of the time interval t. The number of radioactive atoms in the sample at the beginning of the time interval N_t is represented by the variable N, while the variable N1 represents N_{t+dt}, the number of radioactive atoms in the sample at the end of the time increment. Equation 4.23 is applied in line number 1020. The values of N and t are scaled and plotted as the subroutine at line number 3000 is executed. The resulting figure is a graph of the number of atoms remaining in the sample as a function of time.

```
 90 REM              *****   Program 4.3   *****
 95 REM                   Radioactive Decay
100 REM
           *****  set up graphics characteristics  *****

110 SCREEN 2 : CLS : XO = 320 : YO = 100 : SX = 1.5 : SY = SX/2.25
150 REM
              *****  specify initial conditions  *****
160 N = 100
170 A = .0133 : REM Activity constant for radon-220
180 DT = .1
300 REM
              *****  set up screen display  *****

310 Y1 = 0 : REM draw horizontal axis
320 FOR X1 = -110 TO 110 STEP 2
330 XS = XO + SX*X1 : YS = YO - SY*Y1 : PSET (XS,YS)
340 NEXT X1
350 X1 = 0 : REM draw vertical axis
360 FOR Y1 = -100 TO 100 STEP 1.5
370 XS = XO + SX*X1 : YS = YO - SY*Y1 : PSET (XS,YS)
380 NEXT Y1
390 REM draw coordinate grid
400 FOR X1 = -100 TO 100 STEP 10
410 FOR Y1 = -90 TO 90 STEP 10
420 XS = XO + SX*X1 : YS = YO - SY*Y1 : PSET (XS,YS)
430 NEXT Y1
440 NEXT X1
450 SC = 10: REM scale for screen grid in meters
460 SX = 10*SX/SC : SY = SX/2.25
470 LOCATE 1,55 : PRINT "one unit =";SC;"s"
1000 REM
              *****  calculations and plotting  *****

1010 FOR T = 0 TO 156 STEP DT
1020 N1 = N -N*A*DT
1030 GOSUB 3000
1040 N = N1
1050 NEXT T
1060 END
2990 REM
              *****  transformation subroutine  *****
3000 XS = XO +SX*T : YS = YO - SY*N : PSET (XS,YS)
3010 RETURN
```

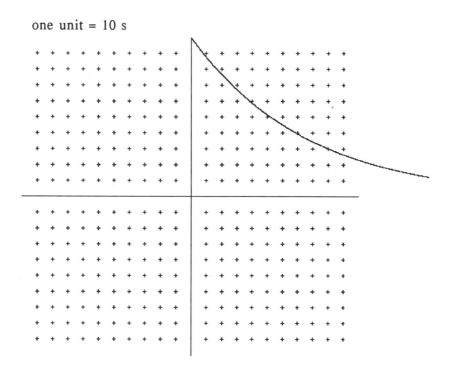

Figure 4.2 The number of remaining nuclei is plotted as function of time using Program 4.3.

4.6 Deflection of Electrons in a Uniform Electric Field

The discovery of cathode rays provided one of the great scientific mysteries of the nineteenth century. Early experimenters such as J. J. Thomson devised many interesting and original experiments in an effort to understand the nature of these electrically charged beams. Cathode rays are produced in a partial vacuum when a large voltage is applied to electrodes enclosed in glass tubes. Experimenters verified that the mysterious cathode rays were emitted from the negative electrode (called the cathode) and travelled to the positive electrode (called the anode).

Cathode rays are now known to be beams of electrons. Electron beams can travel unimpeded through vacuum chambers, and their paths can be altered by the influence of electric and magnetic fields. For example, beams of electrons are deflected by electric fields in oscilloscopes, while the CRTs (cathode ray tubes) of television sets and computer monitors produce images using beams of electrons deflected by magnetic fields. The motion of electrons travelling through an electric field applied at right angles to the original path of the electrons is simulated using Program 4.3.

Electrons originally move parallel to the horizontal (x) axis of the graphic template. A voltage V is applied between two metal plates located a distance d apart and placed above and below the electron beam parallel to the original path of the electrons. In the figure created by the computer program, the metal plates are located parallel to the upper and lower edges of the graphic template. The bottom plate is connected to the positive terminal of the voltage source. The electric field between the parallel plates is uniform and parallel to the y-axis with a magnitude given by Eq 4.24.

$$E = V/d \qquad (4.24)$$

The magnitude of the force affecting the motion of the electron is determined using Eq 4.25. The force is directed parallel to the y-axis.

$$F_y = eE \qquad (4.25)$$

Applying Newton's second law, the equation of motion for a moving electron is

$$eE = md^2y/dt^2 \qquad (4.26)$$

The last-point approximation defined by Eq 4.11 and Eq 4.12 is used to evaluate Eq 4.26 and thus determine the vertical component of the speed and the y-coordinate of the position of the moving electron as a function of time.

$$dy/dt_{t+dt} = dy/dt_t + (e/m)E_y dt \qquad (4.27)$$

$$y_{t+dt} = y_t + (dy/dt_{t+dt})dt \qquad (4.28)$$

Because the electric field is applied perpendicular to the horizontal axis, the horizontal component of force is zero. The horizontal component of velocity v_x is thus constant. The x-coordinate of the position of the electron is determined using Eq 4.29 where x_0 is the x-coordinate of the location of the electron when $t = 0$.

$$x_t = x_0 + v_x t \qquad (4.29)$$

In Program 4.4, the last-point approximation is used to determine the location and speed of the electron. The applied voltage is specified by the variable V in line number 180 and the distance between the metal plates is specified by the variable D in line number 190. The strength of the applied electric field designated by the variable EY is calculated in line number 200 using Eq 4.24. The initial position of the moving electron is designated by the variables X0 and Y in line number 210. The initial components of velocity are designated by the variables VX and VY in line number 220.

Equations 4.27 and 4.28 are evaluated in line number 1030 and line number 1040, respectively. Equation 4.29 is evaluated in line number 1020. The values of X and Y, the coordinates of the location of the electron, are scaled and plotted in the subroutine beginning at line number 3000.

```
 90 REM              *****  Program 4.4  *****
 95 REM   Deflection of an Electron in an Electric Field
100 REM
           *****  set up graphics characteristics  *****
110 SCREEN 2 : CLS : XO = 320 : YO = 100 : SX = 1.5 : SY = SX/2.25
```

4.6 DEFLECTION OF ELECTRON IN E FIELD

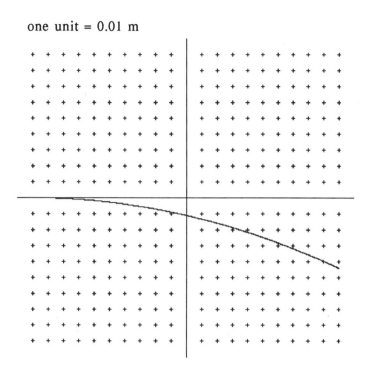

Figure 4.3 A moving electron is deflected by an electric field.

```
150 REM
          *****  specify initial conditions  *****

160 Q = -1.6*10^-19
170 M = 9.100001*10^-31
180 V = 10 : REM Deflector Voltage
190 D = .2
200 EY = V/D
210 X0 = -.1 :Y = 0
220 VX = 2*10^6 : VY = 0
230 DT = 10^-10
300 REM
          *****  set up screen display  *****

310 Y1 = 0 : REM draw horizontal axis
320 FOR X1 = -110 TO 110 STEP 2
330 XS = XO + SX*X1 : YS = YO - SY*Y1 : PSET (XS,YS)
340 NEXT X1
350 X1 = 0 : REM draw vertical axis
360 FOR Y1 = -100 TO 100 STEP 1.5
370 XS = XO + SX*X1 : YS = YO - SY*Y1 : PSET (XS,YS)
380 NEXT Y1
390 REM draw coordinate grid
400 FOR X1 = -100 TO 100 STEP 10
```

```
410 FOR Y1 = -90 TO 90 STEP 10
420 XS = XO + SX*X1
430 NEXT Y1
440 NEXT X1
450 SC = .01 : REM scale for screen grid in meters
460 SX = 10*SX/SC : SY = SX/2.25
470 LOCATE 1,55 : PRINT "one unit =";SC;"m"
1000 REM
          *****  calculations and plotting  *****

1010 FOR T = 0 TO 10^-7 STEP DT
1020 X = XO + VX*T
1030 V1 = VY + (Q*EY/M)*DT
1040 Y1 = Y + V1*DT
1050 GOSUB 3000
1060 VY = V1 : Y = Y1
1070 NEXT T
2000 END
2990 REM
          *****  transformation subroutine  *****

3000 XS = XO + SX*X : YS = YO - SY*Y : PSET (XS,YS)
3010 RETURN
```

4.7 Deflection of Electrons in a Uniform Magnetic Field

The Lorentz force affects moving electric charges passing through a magnetic field. Because of the Lorentz force, an electron beam is deflected as the beam passes through a magnetic field. This fact was used by Jean Perrin in 1895 to show that cathode rays were in fact beams of negatively charged particles. Quantitative measurements of the deflection were later used by J. J. Thomson to determine e/m, the charge-to-mass ratio for the electron.

The magnitude and direction of the Lorentz force acting on a moving charge moving with a velocity v relative to a magnetic field of strength B is calculated using Eq 4.30.

$$F = q(v \times B) \tag{4.30}$$

Program 4.5 simulates the motion of an electron moving through a magnetic field B directed out of the computer screen. (This is the $+z$-direction relative to the graphic template.) When expressed in cartesian coordinates, the components of the Lorentz force affecting the motion of the electron are found to be

$$F_x = qv_yB \tag{4.31}$$

$$F_y = qv_xB \tag{4.32}$$

$$F_z = 0 \tag{4.33}$$

4.7 DEFLECTION OF ELECTRON IN B FIELD

Newton's second law yields a set of three coupled second-order differential equations, one equation for each component of the motion.

$$d^2x/dt^2 = qv_yB/m \qquad (4.34)$$

$$d^2y/dt^2 = qv_xB/m \qquad (4.35)$$

$$d^2z/dt^2 = 0 \qquad (4.36)$$

If the initial position of the moving electron is in the plane of the graphic template and the initial component of velocity perpendicular to the computer screen (the z-component of velocity) is zero, Eq 4.36 can be ignored. This restriction limits the motion of the moving electron to the plane of the graphic template. The last-point approximation is applied in Program 4.5 to evaluate Eq 4.34 and Eq 4.35. Applying Eq 4.11 and Eq 4.12, the set of equations becomes

$$dx/dt_{t+dt} = dx/dt_t + (q/m)v_yBdt \qquad (4.37)$$

$$dy/dt_{t+dt} = dy/dt_t + (q/m)v_xBdt \qquad (4.38)$$

$$x_{t+dt} = x_t + (dx/dt_{t+dt})dt \qquad (4.39)$$

$$y_{t+dt} = y_t + (dy/dt_{t+dt})dt \qquad (4.40)$$

In Program 4.5, the strength of the applied magnetic field is specified by the variable B in line number 180. The coordinates of the initial position of the moving electron are specified in line number 210. The initial values of the horizontal and vertical components of velocity are specified by the variables VX and VY in line number 220.

The components of velocity and position are determined by evaluation of Eq 4.37, Eq 4.38, Eq 4.39, and Eq 4.40 (line numbers 1020, 1040, 1030, and 1050, respectively). For these calculations the horizontal and vertical components of velocity at time t, the beginning of each successive time increment, are specified by the variables VX and VY, respectively. The horizontal and vertical components of the velocity at $t + dt$, the end of the time increment, are similarly specified by the variables V1 and V2. The values of X and Y, the coordinates of the location of the electron, are scaled and plotted in the subroutine beginning at line number 3000.

```
 90 REM            *****  Program 4.5  *****
 95 REM   Deflection of an Electron in a Magnetic Field
100 REM
       *****  set up graphics characteristics  *****

110 SCREEN 2 : CLS : XO = 320 : YO = 100 : SX = 1.5 : SY =
SX/2.25
```

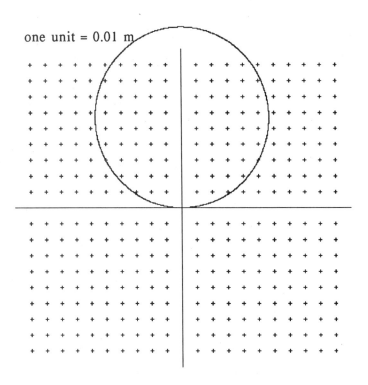

Figure 4.4 A moving electron is deflected by a magnetic field.

```
150 REM
          *****  specify initial conditions  *****

160 Q = -1.6*10^-19
170 M = 9.100001*10^-31
180 B = .0002
210 X=0 :Y = 0
220 VX = 2*10^6 : VY = 0
230 DT = 10^-10
300 REM
          *****  set up screen display  *****

310 Y1 = 0 : REM draw horizontal axis
320 FOR X1 = -110 TO 110 STEP 2
330 XS = XO + SX*X1 : YS = YO - SY*Y1 : PSET (XS,YS)
340 NEXT X1
350 X1 = 0 : REM draw vertical axis
360 FOR Y1 = -100 TO 100 STEP 1.5
370 XS = XO + SX*X1 : YS = YO - SY*Y1 : PSET (XS,YS)
380 NEXT Y1
390 REM draw coordinate grid
400 FOR X1 = -100 TO 100 STEP 10
410 FOR Y1 = -90 TO 90 STEP 10
420 XS = XO + SX*X1 : YS = YO - SY*Y1 : PSET (XS,YS)
```

```
430 NEXT Y1
440 NEXT X1
450 SC = .01 : REM scale for screen grid in meters
460 SX = 10*SX/SC : SY = SX/2.25
470 LOCATE 1,55 : PRINT "one unit =";SC;"m"
1000 REM
     *****  calculations and plotting  *****

1010 FOR T = 0 TO 2*10^-7 STEP DT
1020 V1 = VX +(Q*VY*B/M)*DT
1030 X1 = X + VX*DT
1040 V2 = VY - (Q*VX*B/M)*DT
1050 Y1 = Y +VY*DT
1060 GOSUB 3000
1070 VX = V1 : X = X1
1080 VY = V2 : Y = Y1
1090 NEXT T
2000 END
2990 REM
     *****  transformation subroutine  *****

3000 XS = XO + SX*X : YS = YO - SY*Y : PSET (XS,YS)
3010 RETURN
```

4.8 The Rutherford Scattering Experiment

Because alpha particles emitted from radioactive materials could be detected by flashes of light produced as the alpha particles hit a zinc sulfide screen, Ernest Rutherford designed an experiment to study the scattering of alpha particles by thin metal foils. The experiment was conducted by Geiger and Marsden in 1911 and revealed the existence of the positively charged atomic nucleus.

The imperfect "plum pudding" model which had been described by J. J. Thomson in 1898 proposed that, instead of being concentrated in a compact nucleus, the positive charge occupied most of the volume of the atom. Because the positive charge of the nucleus was not concentrated, an alpha particle passing through a foil of these atoms would be deflected only slightly when an atom was encountered. Large deflections could occur, but only by the cumulative effect of many small deflections. Statistical analysis of this process predicted that the range of scattering angles of the alpha particles should be proportional to the square root of the thickness of the foil through which the alpha particle passed.

It was found instead that alpha particles are often scattered through angles greater than 90 degrees for even the thinnest foils that could be tested. Thorough experiments revealed that the number of alpha particles scattered through large angles is in fact proportional to the thickness of the foil. This result lead Rutherford to propose that the positive charge of the atom is concentrated in a compact nucleus. Because of the Coulomb force, the positively charged alpha particles are repelled by the nucleus and the alpha particles passing close to the nucleus are scattered through large angles.

The magnitude of the force of repulsion between the alpha particle and the nucleus is given by Coulomb's law.

$$F = kQq/r^2 \qquad (4.41)$$

The charge of the atomic nucleus Q is equal to $+Ze$, where Z is the atomic number of the atom. The charge q of the alpha particle is $+2e$. The force on the moving alpha particle is directed away from the nucleus along a line between the alpha particle and the atomic nucleus.

If the path of the alpha particle lies in the *(x-y)* plane, the components of the Coulomb force are given by Eq 4.42 and Eq 4.43, where θ represents the angle between the line to the alpha particle and a line parallel to the horizontal *(x)* axis.

$$F_x = (kQq/r^2)\cos\theta \qquad (4.42)$$

$$F_y = (kQq/r^2)\sin\theta \qquad (4.43)$$

By positioning the coordinate system so that the atomic nucleus is located at the origin of coordinates, $\cos\theta = x/r$ and $\sin\theta = y/r$, where x and y are the coordinates of the position of the alpha particle. In addition, the distance of separation between the nucleus and the alpha particle is $r = (x^2 + y^2)^{1/2}$. Using these relations, the components of the force acting on the alpha particle are

$$F_x = kQqx/r^3 \qquad (4.44)$$

$$F_y = kQqy/r^3 \qquad (4.45)$$

Combining this result with Newton's second law, the equation of motion for the alpha particle yields:

$$d^2x/dt^2 = (kQq/m)x/r^3 \qquad (4.46)$$

$$d^2y/dt^2 = (kQq/m)y/r^3 \qquad (4.47)$$

Program 4.6 applies the last-point approximation to evaluate this set of equations and determine the path of the scattered alpha particle. The mass of the moving alpha particle is specified by the variable M in line number 170. The force constant for Coulomb's law is specified by the variable K in line number 180. The value of Z, the atomic number of the nucleus, is specified in line number 190. Initial values of the horizontal and vertical components of velocity are specified by the variables VX and VY in line number 200 and line number 210. The initial position of the alpha particle is similarly specified in line number 230 and line number 240.

Equations of the form of Eq 4.10 and Eq 4.11 are applied in line number 1020 to line number 1050 to evaluate Eq 4.46 and Eq 4.47. The coordinates X and Y of the position of the alpha particle are scaled and plotted in the subroutine beginning at line number 3000.

4.8 THE RUTHERFORD SCATTERING EXPERIMENT

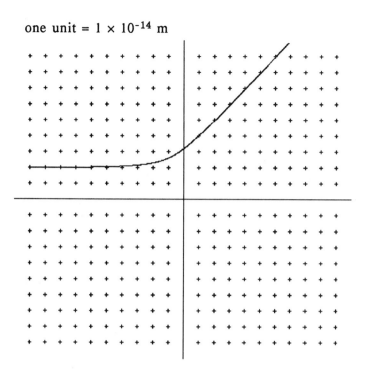

Figure 4.5 An alpha particle is deflected by a gold nucleus.

```
 90 REM            *****  Program 4.6  *****
 95 REM      The Rutherford Scattering Experiment
100 REM
        *****  set up graphics characteristics  *****

110 SCREEN 2 : CLS : XO = 320 : YO = 100 : SX = 1.5 : SY = SX/2.25
150 REM
        *****  specify initial conditions  *****

160 E = 1.6*10^-19
170 M = 4*1.67*10^-27
180 K = 9*10^9
190 Z = 53
200 VX = 1.4*10^7
210 VY = 0
220 X = -10*10^-14
230 Y = 2*10^-14
240 DT = 10^ -22
300 REM
        *****  set up screen display  *****

310 Y1 = 0 : REM draw horizontal axis
320 FOR X1 = -110 TO 110 STEP 2
```

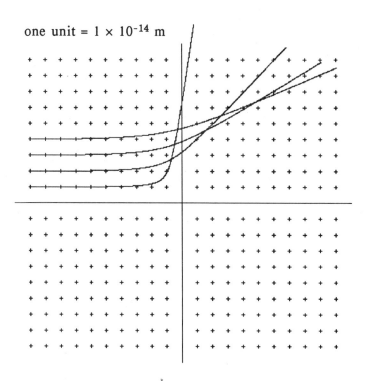

Figure 4.6 Alpha particles are deflected by a gold nucleus.

```
330 XS = XO + SX*X1 : YS = YO - SY*Y1 : PSET (XS,YS)
340 NEXT X1
350 X1 = 0 : REM draw vertical axis
360 FOR Y1 = -100 TO 100 STEP 1.5
370 XS = XO + SX*X1 : YS = YO - SY*Y1 : PSET (XS,YS)
380 NEXT Y1
390 REM draw coordinate grid
400 FOR X1 = -100 TO 100 STEP 10
410 FOR Y1 = -90 TO 90 STEP 10
420 XS = XO + SX*X1 : YS = YO - SY*Y1 : PSET (XS,YS)
430 NEXT Y1
440 NEXT X1
1000 REM     ***** calculations and plotting *****
1002 SC = 10^-14
1004 SX = 10*SX/SC : SY = SX/2.25
1006 LOCATE 1,55 : PRINT " one unit = ";SC;"m"
1010 FOR T = 0 TO 2*10^-20 STEP DT
1020 V1 = VX +((K*Z*E*E/M)/(X^2 + Y^2))*(X/(X^2 + Y^2)^.5)*DT
1030 X1 = X + VX*DT
1040 V2 = VY +((K*Z*E*E/M)/(X^2 + Y^2))*(Y/(X^2 + Y^2)^.5)*DT
1050 Y1 = Y + VY*DT
1060 GOSUB 3000
```

```
1070 VX = V1 : X = X1
1080 VY = V2 : Y = Y1
1090 NEXT T
1200 END
2990
         *****  plotting subroutine  *****
3000 XS = XO + SX*X : YS = YO - SY*Y : PSET (XS,YS)
3010 RETURN
```

Exercises

4.1 In the absence of an electric field, the terminal velocity of a falling oil droplet occurs when the downward gravitational force is balanced by the upward force of fluid friction.

$$(m - m_a)g = bv \tag{4.48}$$

a. Using the variables specified for variables in Program 4.2, calculate the terminal velocity of the oil droplet.
b. Determine the terminal velocity of the oil droplet by running Program 4.1 and compare the result to the calculation in (a).

4.2 a. Alter Program 4.2 to increase the mass of the oil droplet.
b. Alter the program to change the initial position and the initial velocity of the oil droplet.

4.3 a. Alter the equation of motion of the falling oil droplet to include the effect of a vertically directed electric field acting on a charged droplet.
b. Alter Program 4.2 to evaluate this equation of motion.
c. Adjust the value specified for the electric field strength so that the oil droplet is suspended. In this case, the upward electric force just balances the downward gravitational force.

$$(m - m_a)g = qE \tag{4.49}$$

4.4 a. Alter Program 4.4 to double the initial number of radioactive nuclei in the sample.
b. Alter the program to double the value of the activity constant.
c. Alter the program to produce a family of curves with each curve having an incrementally larger activity constant.

4.5 By direct integration of Eq 4.21, the number of atoms remaining in the sample of radioactive atoms is found to be

$$N = N_0 e^{-At} \tag{4.50}$$

a. The half-life of the sample is defined as the time required for one-half of the radioactive nuclei to decay. Using Eq 4.50, show that the half-life is $T_{1/2} = ln(2)/A$.

b. Using Program 4.3 to verify that one-half of the sample decays in the time predicted in (a).

4.6 a. Alter Program 4.4 to simulate the motion of a proton in an electric field.

b. Alter Program 4.4 to produce a family of curves to represent the paths of electrons with incrementally increasing initial velocities.

4.7 The deflection of an electron moving through an electric field is found to be

$$y = (eE/2m)(x^2/v_x^2) \tag{4.51}$$

Verify that the results of Program 4.4 are consistent with Eq 4.51.

4.8 a. Alter Program 4.5 to illustrate the path of a proton moving through a magnetic field perpendicular to the plane of the path of the proton.

b. Increase the value of the magnetic field strength so that the radius of the path of the proton is the same as the radius of the path of an electron moving with the same initial velocity.

4.9 The radius of the path of an electron deflected by a magnetic field is found to be

$$r = Em/eB^2 \tag{4.52}$$

Verify that the results of Program 4.5 are consistent with Eq 4.52.

4.10 a. Alter Program 4.6 to illustrate the paths of scattered protons instead of scattered alpha particles.

b. Alter Program 4.6 to change the nature of the force from repulsive to attractive. Illustrate the paths of particles attracted to the nucleus.

c. By trial and error, find initial conditions such that the particles orbit the nucleus under the influence of the attractive force.

4.11 Alter Program 4.6 to produce Figure 4.6.

Chapter 5

Schrödinger's Equation

5.1 Introduction

The properties of nonrelativistic systems on the atomic scale are consistent with the predictions of Schrödinger's equation. Although many experiments have been conducted to test this idea, no contradictory experimental results have been found.

Schrödinger's equation is a differential equation with solutions which represent the wave functions that characterize the motion of a particle. Although the wave function itself is not directly observable, all of the dynamical variables associated with motion are defined in terms of the wave function. Schrödinger's equation is thus consistent with deBroglie's postulates—motion of particles is linked to properties of the wave function.

The development of Schrödinger's equation is based on fundamental physical principles and provides the basis for the theories of quantum mechanics. Schrödinger's equation reveals that matter possesses properties which are never suspected on the basis of everyday observations. Numerical "experiments" are used in this chapter to investigate phenomena unique to quantum mechanics. Results are depicted graphically on the computer screen.

Although the physical world contains (at least) three spatial dimensions and is certainly time dependent, only the one-dimensional time-independent form of Schrödinger's equation is considered in this text. The properties of matter unique to quantum mechanical systems become apparent even in this restricted case. Thus, complex numerical methods suitable for evaluating Schrödinger's equation for three-dimensional time-dependent systems are unnecessary here. With these restrictions, Schrödinger's equation for an object of mass m under the influence of a conservative force field for which the potential function is $V(x)$ is written as shown in Eq 5.1.

$$-\frac{\hbar^2}{2m}\frac{d^2\psi}{dx^2} + V(x)\psi = E\psi \tag{5.1}$$

In this form, Schrödinger's equation is an ordinary second-order differential equation. The solutions $\psi(x)$ represent the spatial dependence of the wave function of the system under consideration. Solutions exist for any value of the energy of the system E, but only certain of these solutions have characteristics consistent with realistic descriptions of nature. In many cases, nature disallows solutions for all but a few specific values of E. For systems of this type, energy is said to quantized.

The one-dimensional time-independent form of Schrödinger's equation (Eq 5.1) is an example of an eigenvalue equation. Physically allowable solutions for $\psi(x)$, called *eigenfunctions*, are obtained only for certain values of E called the *energy eigenvalues*. Numerical techniques for displaying the eigenfunctions and determining the energy eigenvalues are illustrated in the computer programs of this chapter.

In order to separate the solutions allowed in nature from the infinite number of mathematically allowable solutions of Schrödinger's equation, it is necessary to first interpret the physical significance of the eigenfunction ψ. The wave function incorporates all of the information about the state of a system. The probability interpretation relates the wave function (and thus the eigenfunction) to the values of dynamic variables such as momentum, energy, and position, together with their associated probabilities.

The probability interpretation was proposed by Max Born in 1926. Although the eigenfunction itself is not a physically observable quantity, the square of the eigenfunction is identified as relative probability density: the relative probability per unit length $P(x)$ that the particle is located in a region between x and $x + dx$. The probability (not probability density) that the particle is located in the region between x and $x + dx$ is then:

$$P(x)dx = \psi^2(x)dx \tag{5.2}$$

It is apparent that this interpretation forces several restrictions on the physically allowable solutions of Schrödinger's equation. The allowable mathematical forms of ψ are restricted to well-behaved functions. In addition, because the total probability of locating the particle somewhere must equal unity, it must be possible to normalize the probability density function. In order to meet this requirement, the probability density for particles whose spatial locations are restricted (this situation is called a *bound state*) must approach zero for spatial locations approaching infinity.

Parity is another important characteristic of the eigenfunctions of Schrödinger's equation. In many important cases, the potential energy of the system is represented by a symmetric function.

$$V(x) = V(-x) \tag{5.3}$$

In these cases, if $\psi(x)$ is a solution of Schrödinger's equation, then $+\psi(-x)$ and $-\psi(-x)$ are also solutions. This can be proved by writing Schrödinger's equation with x replaced everywhere by $-x$.

5.2 THE LAST-POINT APPROXIMATION

$$-\frac{\hbar^2}{2m}\frac{d^2\psi(-x)}{dx^2} + V(-x)\psi(-x) = E\psi(-x) \qquad (5.4)$$

If the potential is symmetric, then $V(-x)$ can be replaced by $V(x)$.

$$-\frac{\hbar^2}{2m}\frac{d^2\psi(-x)}{dx^2} + V(x)\psi(-x) = E\psi(-x) \qquad (5.5)$$

Equation 5.5 is identical to Eq 5.1 except that $\psi(x)$ is replaced by $\psi(-x)$ so that both $\psi(-x)$ and $\psi(x)$ are solutions of the same differential equation. Thus the functions can differ only by a multiplicative constant. That is,

$$\psi(-x) = a\psi(x) \qquad (5.6)$$

Because the probability density calculated using either function must satisfy the same normalization condition, the only possible values of the multiplicative constant a which will not alter the normalization of the functions are +1 and -1.

If the value of a is +1, then $\psi(-x) = \psi(x)$ and the eigenfunction is said to have even parity. If the value of a is -1, then $\psi(-x) = -\psi(x)$ and the eigenfunction is said to have odd parity. *Thus, if the potential function $V(x)$ is an even function, all allowable solutions of Schrödinger's equation must have either even parity or odd parity.*

Both the concept of parity and the fact that the probability density approaches zero as the spatial coordinate approaches infinity must be considered in the development of computer programs which evaluate Schrödinger's equation. Parity considerations yield a simple method for finding initial values to start numerical calculations. The fact that the probability density must approach zero as the spatial coordinates approach infinity is used to identify energy eigenvalues of bound particles.

5.2 Schrödinger's Equation and the Last-Point Approximation

To develop a computer program which evaluates the time-independent form of Schrödinger's equation using the last-point approximation, Eq 5.1 is first rewritten as

$$d^2\psi/dx^2 = (2m/\hbar^2)[V(x) - E]\psi \qquad (5.7)$$

Applying the last-point approximation in the form of Eq. 4.10 and Eq 4.11, Eq 5.7 becomes

$$d\psi/dx_{x+dx} = d\psi/dx_x + (2m/\hbar^2)[V(x) - E]\psi dx \qquad (5.8)$$

$$\psi_{x+dx} = \psi_x + d\psi/dx_{x+dx}\, dx \qquad (5.9)$$

In Eq 5.8 and Eq 5.9, the independent variable t of Eq 4.10 and Eq 4.11 has been replaced by the spatial coordinate x, and the dependent variable x has been replaced by the value of the eigenfunction ψ.

In the programs of Chapter 4, initial values of position x and velocity v at time $t = 0$ were specified to start the numerical calculations. A similar procedure is necessary to supply initial values to start the calculations based on Eq 5.8 and Eq 5.9. In this case, values of ψ and $d\psi/dx$ are specified for the location $x = 0$. Correctly chosen, these initial values determine the parity of the solution. This method for finding initial values is subject to the limitation that the potential function $V(x)$ must be a symmetric function.

The slope of a function with even parity (such as $\psi = \cos x$) is zero at the origin. Thus for an eigenfunction with even parity, the initial value chosen for $d\psi/dx$ at the origin is zero and ψ must have an initial value other than zero at the origin. The initial value chosen for ψ determines the amplitude of the eigenfunction.

On the other hand, a function with odd parity (such as $\psi = \sin x$) has a value of zero at the origin. To evaluate an eigenfunction with odd parity, the initial value chosen for ψ at the origin is zero, and $d\psi/dx$ must have an initial value other than zero at the origin. In this case, the initial value chosen for $d\psi/dx$ determines the amplitude of the eigenfunction. (To verify these relations, see Exercise 5.3.)

5.3 One-Dimensional Square Well Potential

The square well potential is a particularly simple case to which the techniques described in the previous section can be applied. Many of the properties of matter revealed by quantum mechanics are illustrated even in this simple example. In addition, numerical results can easily be compared with the predictions of analytical methods. (See Exercise 5.4.) The computer program developed in this section can be extended and applied in other more realistic situations.

For a free particle confined to a one-dimensional square well, the particle is imagined to have zero potential energy in a region of length L occupying a spatially symmetric region from $-L/2 < x < L/2$. Outside this region, the potential energy of the particle is V_0. The total energy E of the particle is less than the amplitude of the potential well V_0. Because $E < V_0$, the particle does not have enough energy to escape and is bound to the potential well.

The potential energy of the system is thus represented by a symmetric function and the eigenfunctions will have either even parity or odd parity. In addition, because the particle is bound to the potential well in the region between $-L/2 < x < L/2$, the amplitude of acceptable eigenfunctions must approach zero outside this region.

Program 5.1 evaluates Eq 5.8 and Eq 5.9 and graphs the result for an electron interacting with a square well potential. The depth of the potential well is -10 eV ($-10 \times 1.6 \times 10^{-19}$ J) and is specified by the variable V in line number 1020. The mass of the electron M is specified in line number 160. The total energy of the electron E is specified to be -3.37 eV ($-3.37 \times 1.6 \times 10^{-19}$ J) in line number 180. (The kinetic energy of the electron is thus 6.63 eV.) The initial values of ψ (Y in program) and $d\psi/dx$ (D in program) are specified in line number 190 and line number 200, respectively. In this example, the initial conditions are those that generate an eigenfunction with even parity. (See Exercise 5.3.)

Equations 5.8 and 5.9 are evaluated for the region from $x = 0$ to the edge of the potential well by execution of the for-next loop beginning at line number 1010. The width of the well is considered to be 1.0 nm (1.0×10^{-9} m), the range of the variable x specified in line number 1010. Over this range, the value of potential energy is $V = -10$ eV. Following execution of this for-next loop, a subsequent for-next loop begins at line number 1110. The value assigned for the potential energy is $V = 0$ eV in this region. The program graphs the eigenfunction from the center of the potential well at the center of the computer screen to the edge of the computer screen. The graphic scale factor declared in line number 450 defines the scale of the figure so that the edge

5. ONE-DIMENSIONAL SQUARE WELL POTENTIAL

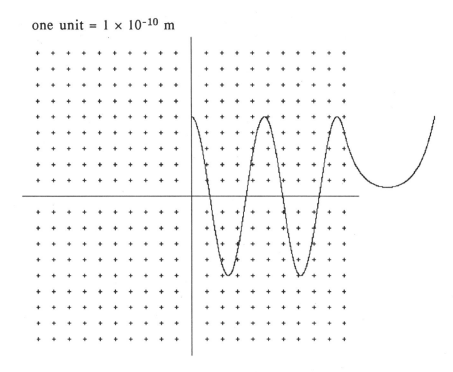

Figure 5.1 A diverging solution for a square potential well does not represent an eigenfunction.

of the potential well is located at the edge of the graphic template. Because the solution has even parity, the graph would be symmetric if it were extended from the center of the screen in the $-x$ direction. (See Exercise 5.8.)

Schrödinger's equation in the form of Eq 5.8 is evaluated in line number 1030. In order to relate the computer program as closely as possible to the mathematical form of Schrödinger's equation, no new variables or reduced variables were introduced in the program. Because the allowable magnitudes of numbers used in many versions of BASIC are restricted to values between 1×10^{38} and 3×10^{-38}, program lines must be written so that multiplication and division operations are ordered to restrict the magnitudes of products and quotients. If the magnitudes of the numbers exceed the limits of the programming language, overflow errors are produced.

Line number 1030 is algebraically equivalent to Schrödinger's equation but variables are grouped so that the magnitude of the numbers encountered in calculations does not exceed the limits of the most common versions of BASIC. Alternatively, advanced versions of BASIC which allow a greater range of numbers can be used or Schrödinger's equation can be written in terms of reduced variables. (For example, let $\hbar = m = 1$.)

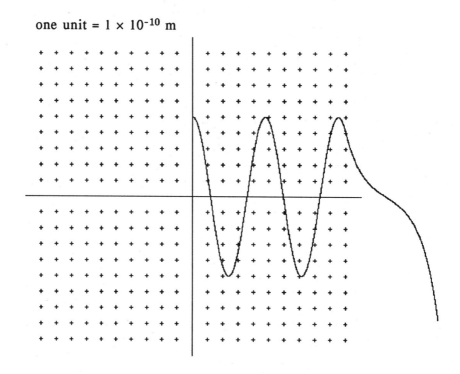

one unit = 1 × 10⁻¹⁰ m

Figure 5.2 A diverging solution for a square potential well does not represent an eigenfunction.

```
 90 REM              *****  Program 5.1  *****
 95 REM                 Square Well Potential
100 REM
         *****  set up graphics characteristics  *****

110 SCREEN 2 : CLS : XO = 320 : YO = 100 : SX = 1.5 : SY = SX/2.25
150 REM      *****  specify initial conditions  *****

160 M = 9.100001*10^-31:REM mass of electron
170 HB = 1.055*10^-34: REM h/2*pi
180 E = -3.37* 1.6*10^-19:REM energy of electron
190 Y = 1:REM amplitude of eigenfunction with even parity
200 D = 0: REM slope of eigenfunction with even parity
210 DX = 10^-12
300 REM
            *****  set up screen display  *****

310 Y1 = 0 : REM draw horizontal axis
320 FOR X1 = -110 TO 110 STEP 2
330 XS = XO + SX*X1 : YS = YO - SY*Y1 : PSET (XS,YS)
340 NEXT X1
```

5. ONE-DIMENSIONAL SQUARE WELL POTENTIAL

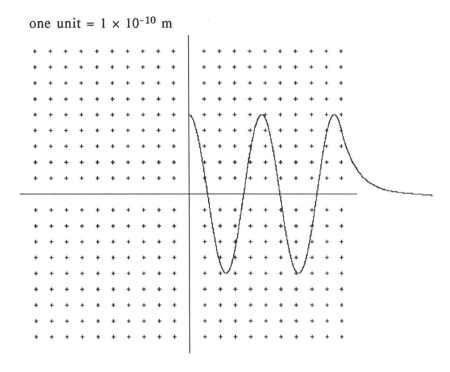

Figure 5.3 A satisfactory eigenfunction for a square potential well. This eigenfunction was found by adjusting the energy E.

```
350  X1 = 0 : REM draw vertical axis
360  FOR Y1 = -100 TO 100 STEP 1.5
370  XS = XO + SX*X1 : YS = YO - SY*Y1 : PSET (XS,YS)
380  NEXT Y1
390  REM draw coordinate grid
400  FOR X1 = -100 TO 100 STEP 10
410  FOR Y1 = -90 TO 90 STEP 10
420  XS = XO + SX*X1 : YS = YO - SY*Y1 : PSET (XS,YS)
430  NEXT Y1
440  NEXT X1
450  SC=10^-10
460  SX = 10*SX/SC : SY = 100\2.25
470  LOCATE 1,55 : PRINT  "one unit =";SC;"m"
1000 REM
         *****   calculations and plotting   *****

1010 FOR X = O TO 10*10^-10 STEP DX
1020 V = -10 * 1.6*10^-19: REM potential function
1030 DY = D + (2*M/HB)*((V - E)*Y/HB)*DX
1040 Y1 = Y +DY*DX
1050 GOSUB 3000
1060 D = DY : Y = Y1
```

one unit = 1 × 10⁻¹⁰ m

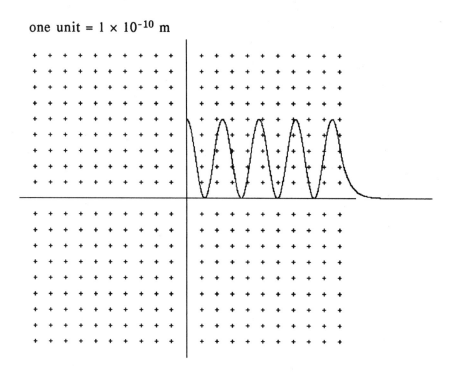

Figure 5.4 This figure depicts the probability density corresponding to eigenfunction of Figure 5.3.

```
1070 NEXT X
1110 FOR X = 10*10^-10 TO 20*10^-10 STEP DX
1120 V = 0 : REM potential function
1130 DY = D + (2*M/HB)*((V - E)*Y/HB)*DX
1140 Y1 = Y +DY*DX
1150 GOSUB 3000
1160 D = DY : Y = Y1
1170 NEXT X
1200 END
2990 REM      ***** plotting subroutine  *****
3000 XS = XO + SX*X : YS = YO - SY*Y : PSET (XS,YS)
3010 RETURN
```

Figure 5.1 displays the results obtained by execution of Program 5.1. The function plotted does not approach zero as x approaches infinity and cannot be normalized. This solution is thus not an eigenfunction. To generate Figure 5.3, the value of the total energy E in line number 180 is changed to -3.3682 eV and the solution behaves satisfactorily, smoothly approaching zero as x increases. Thus -3.3682 eV is an energy eigenvalue for this system. If the energy is increased further to -3.3679 eV, Figure 5.2 results. Again, the function displayed does not approach zero as x approaches infinity and does not represent an eigenfunction. Other values of E can be tested to determine if other energy eigenvalues exist for this system. (See Exercise 5.2.)

To find eigenfunctions which have odd parity and to determine the associated energy eigenvalues, the initial values of ψ and $d\psi/dx$ in line number 190 and line number 200 must be altered. (See Exercise 5.3.) Inside the potential well the eigenfunction ψ is sinusoidal in appearance. The wavelength in this region is in agreement with the deBroglie relation. (See Exercise 5.1.)

Instead of graphing the eigenfunction ψ, the program can be altered to plot the probability density ψ^2. (See Exercise 5.6.) Figure 5.4 displays the probability density of the eigenfunction found in Figure 5.3. Notice that the probability density does not drop to zero at the edge of the potential well but asymptotically approaches zero in the region outside the well. This region is called the classically forbidden region because, according to laws of classical mechanics, the energy of the electron is insufficient to propel the electron into this region. (See Exercise 5.7.) The finite value of the probability density indicates that the particle is not excluded from the classically forbidden region but instead has some probability of being located in this region.

5.4 The Harmonic Oscillator

The harmonic oscillator provides a further example of the application of Schrödinger's equation. A harmonic oscillator consists of an object of mass m under the influence of a linear restoring force $F = -kx$. The potential energy of the system is thus

$$V = kx^2/2 \tag{5.10}$$

The harmonic oscillator furnishes a basis for understanding more complex systems because any realistic potential can be approximated by the harmonic oscillator potential for small oscillations of a particle about the minimum of the potential function. For this potential function, Schrödinger's equation can be written in the form of Eq 5.7.

$$d^2\psi/dx^2 = (2m/\hbar^2)(kx^2/2 - E)\psi \tag{5.11}$$

When the last-point approximation is applied, Eq 5.8 and Eq 5.9 become

$$d\psi/dx_{x+dx} = d\psi/dx_x + (2m/\hbar^2)(kx^2/2 - E)\psi dx \tag{5.12}$$

$$\psi_{x+dx} = \psi_x + d\psi/dx_{x+dx}\, dx \tag{5.13}$$

Program 5.2 evaluates Eq 5.12 and Eq 5.13 and graphs the results. An electron of mass M specified in line number 160 interacts with a linear restoring force with a force constant K specified in line number 210. The energy E of the electron is 3.80155 eV ($3.80155 \times 1.6 \times 10^{-19}$ J) specified in line number 180. The initial values of ψ (Y in program) and $d\psi/dx$ (D in program) are specified in line number 190 and line number 200 to create a solution with odd parity.

Equations 5.12 and 5.13 are evaluated over the region from the center to the edge of the computer screen by the execution of the for-next loop beginning at line number 1010. Equation 5.12 is evaluated in line number 1030. As described in the previous section, multiplication and division operations are ordered to prevent overflow errors.

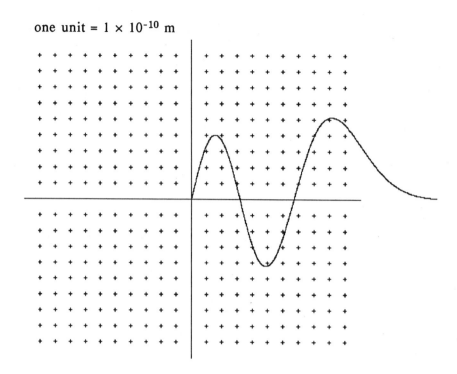

Figure 5.5 A satisfactory eigenfunction for the harmonic oscillator potential.

Because the solution has odd parity, the graph would be inverted if it were extended from the center of the screen in the $-x$ direction. (See Exercise 5.8.)

```
 90 REM              *****  Program 5.2  *****
 95 REM              The Harmonic Oscillator
100 REM
       *****  set up graphics characteristics  *****

110 SCREEN 2 : CLS : XO = 320 : YO = 100 : SX = 1.5 : SY = SX/2.25
150 REM
            *****  specify initial conditions  *****

160 M = 9.100001*10^-31:REM mass of electron
170 HB = 1.055*10^-34: REM h/2*pi
180 E =3.80155* 1.6*10^-19:REM energy of electron
190 Y = 0:REM initial amplitude for eigenfunction with odd parity
200 D = .8*10^10: REM initial slope of eigenfunction with odd parity
210 K=1
220 DX = 10^-12
```

5.4 THE HARMONIC OSCILLATOR

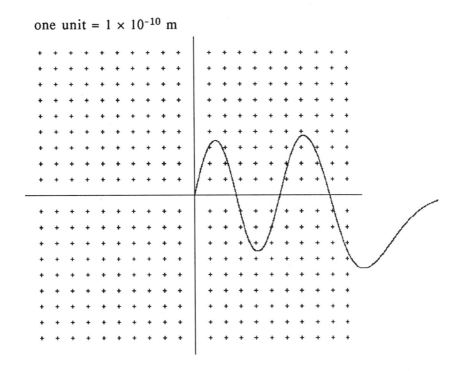

Figure 5.6 A satisfactory eigenfunction for the harmonic oscillator potential. The energy E was increased to find this eigenfunction.

```
300 REM
            *****  set up screen display  *****

310 Y1 = 0 : REM draw horizontal axis
320 FOR X1 = -110 TO 110 STEP 2
330 XS = XO + SX*X1 : YS = YO - SY*Y1 : PSET (XS,YS)
340 NEXT X1
350 X1 = 0 : REM draw vertical axis
360 FOR Y1 = -100 TO 100 STEP 1.5
370 XS = XO + SX*X1 : YS = YO - SY*Y1 : PSET (XS,YS)
380 NEXT Y1
390 REM draw coordinate grid
400 FOR X1 = -100 TO 100 STEP 10
410 FOR Y1 = -90 TO 90 STEP 10
420 XS = XO + SX*X1 : YS = YO - SY*Y1 : PSET (XS,YS)
430 NEXT Y1
440 NEXT X1
450 SC=10^-10
460 SX = 10*SX/SC : SY = 100\2.25
470 LOCATE 1,55 : PRINT  "one unit =";SC;"m"
1000 REM
            *****  calculations and plotting  *****
```

```
1010 FOR X = O TO 20*10^-10 STEP DX
1020 V = .5*K*X*X   : REM potential function
1030 DY = D + (2*M/HB)*((V - E)*Y/HB)*DX
1040 Y1 = Y +DY*DX
1050 GOSUB 3000
1060 D = DY : Y = Y1
1070 NEXT X
1200 END
2990 REM
              ***** plotting subroutine  *****

3000 XS = XO + SX*X : YS = YO - SY*Y : PSET (XS,YS)
3010 RETURN
```

Figure 5.5 displays the results obtained by execution of Program 5.2. The wavelength of the eigenfunction is shorter near the center of the potential well where the kinetic energy of the electron is relatively large. In the classically forbidden region the solution approaches zero asymptotically. Figure 5.6 displays the eigenfunction obtained as the energy of the electron is increased to the next higher energy eigenvalue corresponding to an eigenfunction with odd parity. Other evenly spaced energy eigenvalues can be found. (See Exercise 5.9.)

5.5 The Zero Potential

Schrödinger's equation takes its simplest mathematical form in the case of a constant potential $V = V_0$. A particle moving under the influence of such a potential experiences no force and is called a *free particle*.

To predict the quantum mechanical behavior of a free particle it is necessary to develop a solution to Schrödinger's equation.

$$d^2\psi/dx^2 = (2m/\hbar^2)(V_0 - E)\psi \tag{5.14}$$

By direct substitution, it is easy to verify that there is a solution of the form

$$\psi(x) = Ae^{ikx} \; ; \; k = [2m(E - V_0)]^{1/2}/\hbar \tag{5.15}$$

This solution, in fact, represents a free particle moving in the $+x$ direction. Using Euler's identity, the complex exponential of Eq 5.15 can be written

$$\psi(x) = A\cos(kx) + iA\sin(kx) \tag{5.16}$$

It is important to note that $\cos(kx)$ is a function which represents an eigenfunction with even parity and $\sin(kx)$ is a function which represents an eigenfunction with odd parity. Thus the eigenfunction ψ has a part with even parity and a separate part with odd parity. The correct solution is a mixture of these two forms. In addition, the am-

5.5 THE ZERO POTENTIAL

plitude of the eigenfunction A is a complex quantity. The complex number A can be represented as $A = A_1 e^{i\phi}$, where A_1 is a real number. These results are expressed in Eq 5.17.

$$\psi(x) = A_1 e^{i\phi} \cos(kx) + i A_1 e^{i\phi} \sin(kx) \tag{5.17}$$

Defining $\psi_{\text{even}} = A_1 \cos(kx)$ and $\psi_{\text{odd}} = A_1 \sin(kx)$, Eq 5.17 becomes

$$\psi(x) = e^{i\phi}(\psi_{\text{even}} + i\psi_{\text{odd}}) \tag{5.18}$$

The probability density can be calculated using Eq 5.18. In the case of a complex function, the probability density must be a real number and is thus equal to the product of the wave function ψ and its complex conjugate ψ^*.

$$P(x) = \psi^*(x)\psi(x) \tag{5.19}$$

Using this definition with the wave function of Eq 5.18

$$P(x) = [e^{-i\phi}(\psi_{\text{even}} - i\psi_{\text{odd}})][e^{i\phi}(\psi_{\text{even}} + i\psi_{\text{odd}})] \tag{5.20}$$

Equation 5.20 reduces to

$$P(x) = \psi^2_{\text{even}} + \psi^2_{\text{odd}} \tag{5.21}$$

Thus for a free particle, the probability density can be calculated in terms of ψ_{even} and ψ_{odd}. Because the free particle is not constrained to any particular spatial region, it is not possible to normalize the probability density. In practice, free particles are confined to a finite region and the normalization is performed by integrating the probability over the length of the beam to which the particle is confined.

Program 5.3 evaluates Schrödinger's equation for a free particle. Solutions with both even parity and odd parity are evaluated by selecting the initial conditions as described in Section 5.2.

For the free particle, the solution with even parity is represented by $\psi_{\text{even}} = A_1 \cos(kx)$. Consistent with this, the values chosen at $x = 0$ are $\psi_{\text{even}} = 1$ and $d\psi_{\text{even}}/dx = 0$. The solution with odd parity is represented by $\psi_{\text{odd}} = A_1 \sin(kx)$. Thus the slope is at any point is the derivative $d\psi_{\text{odd}}/dx = A_1 k \cos(kx)$. This is equivalent to $k\psi_{\text{even}}$. The correct values at $x = 0$ are thus $\psi_{\text{odd}} = 0$ and $d\psi_{\text{odd}}/dx = k$. Using these initial values to start the numerical calculations, the probability density is calculated and plotted using Eq 5.21.

The mass M of the particle is specified in line number 160. The energy E and the potential energy V of the particle are specified in line number 180 and line number 182. The value of k (the variable K) is determined using Eq 5.15 in line number 185. Boundary conditions appropriate for ψ_{even} and for ψ_{odd} are specified in line number 190 and line number 200, respectively. The variable YR represents ψ_{even} and the variable YI represents ψ_{odd}. Schrödinger's equation in the form of Eq 5.8 and Eq 5.9 is evaluated to determine ψ_{even} in line number 1030 and line number 1040 while ψ_{odd} is similarly determined in line number 1060 and line number 1070. Both ψ_{even} (YR) and ψ_{odd} (YI) are plotted by successive execution of the subroutine beginning at line number 3000. In order to plot YR, the variable Y to be plotted is specified as equal to YR in line number 1040. Similarly, in line number 1070, Y is specified as equal to YI.

The probability density represented by the variable P is calculated in line number 1090 using Eq 5.21. The probability density is also plotted by execution of the subroutine beginning at line number 3000.

The calculations proceed in the $+x$ direction from the origin of coordinates as the for-next loop in line number 1010 is executed. A similar loop beginning in line number 1110 performs calculations in the $-x$ direction from the origin of coordinates.

```
 90 REM              *****  Program 5.3  *****
 95 REM        The Zero Potential (Free Particle)
100 REM
         *****  set up graphics characteristics  *****

110 SCREEN 2 : CLS : XO = 320 : YO = 100 : SX = 1.5 : SY = SX/2.25
150 REM
         *****  specify initial conditions  *****

160 M = 9.100001*10^-31
170 HB = 1.055*10^-34
180 E = 10* 1.6*10^-19
182 V = 0
185 K = SQR(2)*SQR(M)*SQR(E-V)/HB : REM compute value of
wave number k
190 YR = 1 : D1 = 0 : REM initial slope and intercept of
real term
200 YI = 0 : D2 = K : REM initial slope and intercept for
imaginary term
210 DX = 10^-12
300 REM
         *****  set up screen display  *****

310 Y1 = 0 : REM draw horizontal axis
320 FOR X1 = -110 TO 110 STEP 2
330 XS = XO + SX*X1 : YS = YO - SY*Y1 : PSET (XS,YS)
340 NEXT X1
350 X1 = 0 : REM draw vertical axis
360 FOR Y1 = -100 TO 100 STEP 1.5
370 XS = XO + SX*X1 : YS = YO - SY*Y1 : PSET (XS,YS)
380 NEXT Y1
390 REM draw coordinate grid
400 FOR X1 = -100 TO 100 STEP 10
410 FOR Y1 = -90 TO 90 STEP 10
420 XS = XO + SX*X1 : YS = YO - SY*Y1 : PSET (XS,YS)
430 NEXT Y1
440 NEXT X1
```

5.5 THE ZERO POTENTIAL

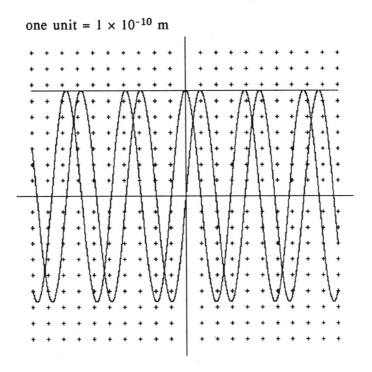

Figure 5.7 Probability density for a free particle is determined using Program 5.3.

```
450 SC = 10^-10
460 SX = 10*SX/SC : SY = 100/2.25
470 LOCATE 1,55 : PRINT "one unit =";SC;"m"
1000 REM
           *****  calculations and plotting  *****

1010 FOR X = 0 TO 10*10^-10 STEP DX
1020 V = 0 : REM potential function
1030 DR = D1+ (2*M/HB)*((V - E)*YR/HB)*DX
1040 YR = YR+DR*DX : Y = YR
1050 GOSUB 3000
1060 DI = D2+ (2*M/HB)*((V - E)*YI/HB)*DX
1070 YI = YI+DI*DX : Y = YI
1080 GOSUB 3000
1090 P = YI^2 + YR^2 : Y = P
1095 GOSUB 3000
1100 D1= DR : D2 = DI
1105 NEXT X
1106 YR = 1 : D1 = 0 : REM initial slope and intercept of
real term
1108 YI = 0 : D2 = K : REM  initial slope and intercept for
imaginary term
1110 FOR X = 0 TO -10*10^-10 STEP -DX
1120 V = 0 : REM potential function
```

```
1130 DR = D1+ (2*M/HB)*((V - E)*YR/HB)*(-DX)
1140 YR = YR+DR*(-DX) : Y = YR
1150 GOSUB 3000
1160 DI = D2+ (2*M/HB)*((V - E)*YI/HB)*(-DX)
1170 YI = YI+DI*(-DX) : Y = YI
1180 GOSUB 3000
1185 P = YI^2 + YR^2 : Y = P
1190 GOSUB 3000
1200 D1= DR : D2 = DI
1205 NEXT X
1300 END
2990 REM
          ***** plotting subroutine *****

3000 XS = XO +SX*X : YS = YO -SY*Y : PSET (XS,YS)
3010 RETURN
```

5.6 The Step Potential

Program 5.3 can be altered to determine how the probability density of the free particle is affected by the presence of a potential step. The potential step occurs when the value of the potential energy V changes abruptly. In the computer program used to produce Figure 5.8, the change in potential energy occurs at the origin of coordinates located at the center of the graphic template. Program 5.3 is modified so that the value of potential energy V specified in line number 182 and in line number 1020 is increased to 5 eV for the $+x$ region but the value of potential energy specified in line number 1120 for the $-x$ region is unaltered. The program proceeds as before. The probability density is first determined for positive values of x. Execution of a second for-next loop beginning at line number 1110 then determines the probability density for negative values of x. The results are illustrated in Figure 5.8. Figure 5.9 illustrates the results of a similar calculation except that the value of the potential energy V in line number 1020 was reduced to -5 eV.

In both Figure 5.8 and Figure 5.9, electrons travelling from the region for which $x < 0$ interact with a potential step located at $x = 0$. The transmission of electrons past the potential barrier is indicated by the uniform probability density in the region for which $x > 0$. Thus, as in the case of free particles, the probability density is constant in the region occupied by electrons transmitted past the potential barrier.

In contrast, the probability density is not uniform in the region $x < 0$. The wavelike variation of the probability density indicates the presence of reflected electrons. In this region, reflected particle waves interfere with the incident particle waves just as sound waves in a pipe reflected from the end of the pipe interfere with incident waves. As in the case of sound waves, interference of the oppositely directed particle waves produces a standing wave. The fraction of incident electrons reflected by the potential barrier can be determined from the amplitude of the variation of the probability density.

The reflection coefficient R is the ratio of the intensity of the reflected wave to the intensity of the incident wave. The eigenfunction ψ_i associated with the incident particle wave is represented by Eq 5.22, and the eigenfunction ψ_r associated with the reflected particle wave is represented by Eq 5.23.

5.6 THE STEP POTENTIAL

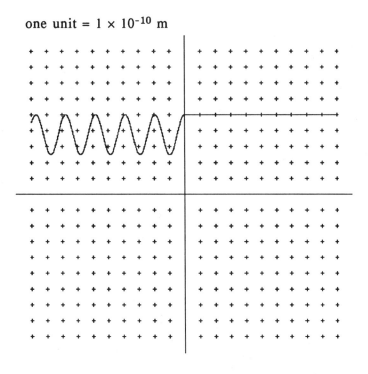

Figure 5.8 The probability density of a particle encountering a potential step is shown. The potential V is zero in the $-x$ region and has the value $V = -5.0$ eV in the $+x$ region.

$$\psi_i = Ae^{ikx} \tag{5.22}$$

$$\psi_r = Be^{-ikx}e^{i\theta} \tag{5.23}$$

Thus the reflection coefficient R is defined as

$$R = \psi^*_r \psi_r / \psi^*_i \psi_i = B^2/A^2 \tag{5.24}$$

In the region for which $x < 0$, the wave function is a linear superposition of the incident wave and the reflected wave. Thus for this region

$$\psi = \psi_i + \psi_r = Ae^{ikx} + Be^{-ikx}e^{i\theta} \tag{5.25}$$

The probability density is then

$$P(x) = \psi^*\psi = (Ae^{-ikx} + Be^{ikx}e^{-i\theta})(Ae^{ikx} + Be^{-ikx}e^{i\theta}) \tag{5.26}$$

$$P(X) = A^2 + B^2 + AB(e^{2ikx}e^{-i\theta} + e^{-2ikx}e^{i\theta}) \tag{5.27}$$

Using Euler's relation, this reduces to

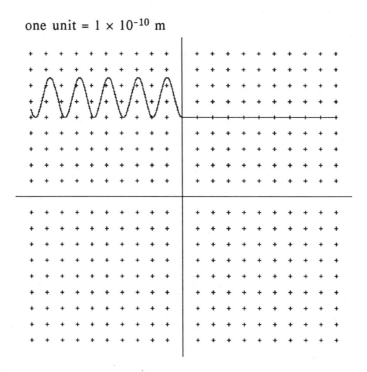

Figure 5.9 The probability density of a particle encountering a potential step is shown. The potential V is zero in the $-x$ region and has the value $V = 5.0$ eV in the $+x$ region.

$$P(x) = A^2 + B^2 + 2AB \cos(2kx - \theta) \tag{5.28}$$

Relating this result to Figure 5.8, we see that the maximum value of the probability density P_{max} occurs when the cosine function in Eq 5.28 has the value +1 and the minimum value P_{min} occurs when the cosine has the value -1. Thus,

$$P(x)_{max} = A^2 + B^2 + 2AB = (A + B)^2 \tag{5.29}$$

$$P(x)_{min} = A^2 + B^2 - 2AB = (A - B)^2 \tag{5.30}$$

This relation allows the reflection coefficient R to be determined from measurements of P_{max} and P_{min} made from the computer screen. From Eq 5.29 and Eq 5.30

$$(P_{min}/P_{max})^{1/2} = (A - B)/(A + B) \tag{5.31}$$

Substituting Eq 5.24, the definition of R,

$$(P_{min}/P_{max})^{1/2} = (1 - R^{1/2})/(1 + R^{1/2}) \tag{5.32}$$

Solving for R,

$$R = [1 - (P_{min}/P_{max})^{1/2}]^2/[1 + (P_{min}/P_{max})^{1/2}]^2 \qquad (5.33)$$

Thus the reflection coefficient can be obtained from figures on the computer screen by measuring the maximum and minimum values of the probability density in the region where reflection occurs. (See James E. Draper, *American Journal of Physics*, Vol. 47, p. 525, 1979.) The reflection coefficient is determined from Eq 5.33. For example, in Figure 5.8, $P_{max} = 13.6$ and $P_{min} = 6.6$. Using these values, the calculated value of the reflection coefficient *is* $R = 0.032$. For Figure 5.9, the reflection coefficient is found to be $R = 0.010$. These values represent the fraction of electrons reflected when an electron beam of energy $E = 10$ eV encounters a potential step with an amplitude of 5 eV.

The transmission coefficient T represents the fraction of electrons transmitted across the potential step. Because in the simple case considered here, the incident particles can only be transmitted or reflected as they interact with the potential barrier, $R + T = 1$. Thus the transmission coefficient can be determined.

$$T = 1 - R \qquad (5.34)$$

These results represent a major change from the predictions of classical mechanics. Classically, all of the electrons would be transmitted past the barrier as long as the energy of the electron E was greater than the potential energy of the barrier. However, quantum mechanics predicts a fraction R of the electrons will be reflected. Experimental results agree with the predictions of quantum mechanics.

5.7 The Barrier Potential

One of the most fascinating predictions of quantum mechanics is the phenomenon of barrier penetration or "tunneling." A potential "barrier" can be represented by a region of width w in which the potential energy V is higher than the potential energy of the surrounding region. Schrödinger's equation correctly predicts that a fraction of the incident particles can pass through this barrier even if the energy of the incident particles is less than the potential energy in the region occupied by the barrier. Classically, when $E < V$, all of the incident particles would be reflected from the barrier, and when $E > V$, all of the particles would be transmitted past the barrier. This system provides a particularly clear example of the wave nature of particles and the statistical nature of the predictions of quantum mechanics.

Program 5.4 simulates the interaction of electrons with a potential barrier. The region of the barrier extends from $x = 0$ to $x = -w$. The potential outside the barrier is zero, and the potential in the region occupied by the barrier is V. The values of barrier width w and barrier height V are specified in line number 155 and line number 1120, respectively. The reflection and transmission coefficients for the barrier can be determined from the drawing created on the computer screen using Eq 5.33 and Eq 5.34.

The program is similar to Program 5.3 except that Schrödinger's equation is evaluated in three distinct regions. The for-next loop beginning at line number 1010 performs calculations to the right of the barrier in the $+x$ region. Solutions with both even parity and odd parity are evaluated, and the probability density represented by P is calculated in line number 1090.

84   5 · SCHRÖDINGER'S EQUATION

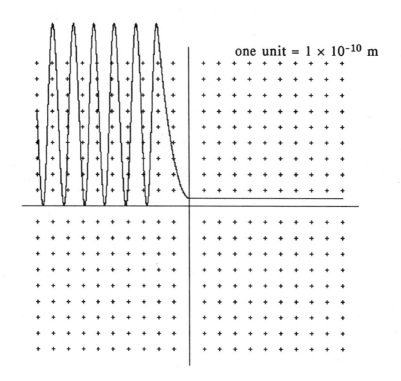

Figure 5.10 This figure represents the probability density of a particle encountering a potential barrier. The barrier thickness w is 2×10^{-10} m.

The for-next loop beginning at line number 1110 then evaluates Schrödinger's equation for the region occupied by the barrier extending from $x = 0$ to $x = -w$. The probability density P is determined in line number 1185. The initial values used for ψ (YR and YI represent ψ_r and ψ_i) and for $d\psi/dx$ (D1 and D2 represent $d\psi_r/dx$ and $d\psi_i/dx$) are those specified in line number 1106 and line number 1108.

The final for-next loop beginning at line number 1110 then performs calculations for the region from the edge of the barrier located at $x = -w$ to the edge of the graphic template. The initial values of ψ and $d\psi/dx$ are those found at $x = -w$ after execution of the preceding for-next loop.

```
 90 REM          *****  Program 5.4  *****
 95 REM       The Barrier Potential (Tunneling)
100 REM
        *****  set up graphics characteristics  *****

110 SCREEN 2 : CLS : XO = 320 : YO = 100 : SX = 1.5 : SY =
SX/2.25
150 REM
           *****  specify initial conditions  *****

155 W = .2*10^-10 : REM width of potential barrier
```

5.7 THE BARRIER POTENTIAL

```
160 M = 9.100001*10^-31
170 HB = 1.055*10^-34
180 E = 10* 1.6*10^-19
182 V = 0 * 1.6*10^-19
185 K = SQR(2)*SQR(M)*SQR(E-V)/HB
190 YR= 1 : D1 = 0 : REM initial intercept and slope for
real term
200 YI =0 : D2 = K : REM initial intercept and slope for
imaginary term
210 DX = 10^-12
300 REM           *****  set up screen display  *****

310 Y1 = 0 : REM draw horizontal axis
320 FOR X1 = -110 TO 110 STEP 2
330 XS = XO + SX*X1 : YS = YO - SY*Y1 : PSET (XS,YS)
340 NEXT X1
350 X1 = 0 : REM draw vertical axis
360 FOR Y1 = -100 TO 100 STEP 1.5
370 XS = XO + SX*X1 : YS = YO - SY*Y1 : PSET (XS,YS)
380 NEXT Y1
390 REM draw coordinate grid
400 FOR X1 = -100 TO 100 STEP 10
410 FOR Y1 = -90 TO 90 STEP 10
420 XS = XO + SX*X1 : YS = YO - SY*Y1 : PSET (XS,YS)
430 NEXT Y1
440 NEXT X1
450 SC = 10^-10
460 SX = 10*SX/SC : SY = 100/2.25
470 LOCATE 1,55 : PRINT "one unit =";SC;"m"
1000 REM    *****  calculations and plotting  *****

1010 FOR X = 0 TO 10*10^-10 STEP DX
1020 V = 0
1030 DR = D1+ (2*M/HB)*((V - E)*YR/HB)*DX
1040 YR = YR+DR*DX
1060 DI = D2+ (2*M/HB)*((V - E)*YI/HB)*DX
1070 YI = YI+DI*DX
1090 P = YI^2 + YR^2 : Y = P
1094 GOSUB 3000
1096 D1= DR : D2 = DI
1100 NEXT X
1106 YR= 1 : D1 = 0 : REM initial slope and intercept of
real term
1108 YI= 0 : D2 = K : REM initial slope and intercept of
imaginary term
1110 FOR X = 0 TO -W STEP -DX
1120 V =   20 *1.6*10^-19 : REM height of potential barrier
```

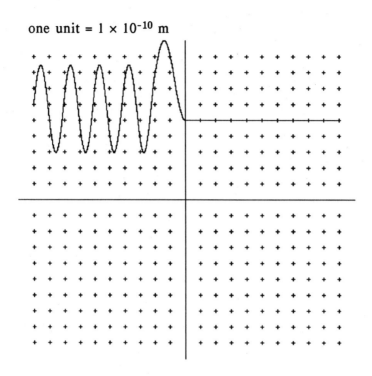

Figure 5.11 In this figure, the energy of the particle E is greater than the potential of the barrier.

```
1130 DR = D1+ (2*M/HB)*((V - E)*YR/HB)*(-DX)
1140 YR = YR+DR*(-DX)
1160 DI = D2+ (2*M/HB)*((V - E)*YI/HB)*(-DX)
1170 YI = YI+DI*(-DX)
1185 P = YI^2 + YR^2 :Y = P
1190 GOSUB 3000
1200 D1= DR : D2 = DI
1205 NEXT X
1300 FOR X = -W TO - 10 *10^-10 STEP -DX
1320 V = 0
1330 DR = D1+ (2*M/HB)*((V - E)*YR/HB)*(-DX)
1340 YR = YR+DR*(-DX)
1360 DI = D2+ (2*M/HB)*((V - E)*YI/HB)*(-DX)
1370 YI = YI+DI*(-DX)
1385 P = YI^2 + YR^2: Y = P
1390 GOSUB 3000
1400 D1= DR : D2 = DI
1405 NEXT X
1500 END
2990 REM     ***** plotting subroutine *****
3000 XS = XO +SX*X : YS = YO -SY*Y : PSET (XS,YS)
3010 RETURN
```

Results of calculations are shown in Figures 5.10 and 5.11. Figure 5.10 depicts the probability density in the case $E < V$. In the region occupied by the barrier, the probability density appears to decay exponentially; however, the particle has a finite probability density on either side of the barrier. Clearly, the particle can "tunnel" through the classically forbidden region. The reflection coefficient is found to be $R = 0.964$ using Eq 5.33. Classically, the expected value of the reflection coefficient is $R = 1$.

In Figure 5.11, the energy of the incident electrons is greater than the potential energy of the barrier. In this case, the laws of classical physics predict that the reflection coefficient will be zero because all electrons have sufficient energy to overcome the barrier. In this figure, the reflection coefficient is instead found to be $R = 0.071$.

5.8 Fourier Transforms and the Uncertainty Principle

The predictions of Schrödinger's equation have revealed that systems on the atomic scale have several startling properties for which everyday experience provides no analogies. For example, rather than making absolute predictions, the wave-like eigenfunctions which are found to be physically allowable solutions to Schrödinger's equation can only predict probabilities. Calculations of probability density reveal that objects can sometimes be found in regions from which particles are excluded according to the laws of classical mechanics. Where classical mechanics allows any value of energy, the energy of a bound particle is instead restricted to a specific set of energy eigenvalues. In addition, particles can be reflected from seemingly transparent potential steps but may pass through classically impenetrable potential barriers.

The purpose of this section is to explore yet another of these subtle properties of matter in the quantum mechanical realm—the idea of uncertainty in the values of dynamical variables. The uncertainty principle was first discovered by Werner Heisenberg in 1927. According to the uncertainty principle, the lowest possible value of the product of the uncertainty of any two conjugate dynamical variables is equal to Planck's constant $\hbar$. Examples of sets of conjugate dynamical variables include position and momentum (p and x) or energy and time (E and t). Thus, the product of the uncertainties dp and dx of the values of the dynamical variables p and x is

$$dxdp > \hbar \quad (5.35)$$

According to the deBroglie relation, the momentum of a particle is $p = \hbar k$. Thus the uncertainty in momentum dp is

$$dp = \hbar dk \quad (5.36)$$

Substituting Eq 5.36 into Eq 5.37, the uncertainty principle becomes

$$dkdx > 1 \quad (5.37)$$

Because k and x are related through the eigenfunctions of Schrödinger's equation, the eigenfunctions can be used to explore and verify this uncertainty relation for the dynamical variables k and x. However, in order to simplify the computer programs, mathematical functions with properties like those of eigenfunctions with even parity will be used in this section instead of actual eigenfunctions determined from numerical solutions of Schrödinger's equation. This technique does not diminish the significance of the results because each of the functions studied would in fact be an eigenfunction of Schrödinger's equation for some undetermined potential function $V(x)$.

An eigenfunction of Schrödinger's equation $\psi(x)$ is called the *position representation* because it discloses the spatial characteristics of the probability density. In addition to representing the eigenfunction as a mathematical function $\psi(x)$, the eigenfunction can also be formed by the superposition of a distribution of sinusoidal functions [$\sin(kx)$ and $\cos(kx)$] with a continuous spectrum of values of k. This representation of the eigenfunction is called the *momentum representation* $\Phi(k)$. The momentum representation of the eigenfunction is equivalent to finding the Fourier transform of the position representation of the eigenfunction.

In order to calculate the product of the uncertainties (in the form of Eq 5.37), the uncertainty of position is determined from the position representation $\psi(x)$ and the uncertainty of momentum is determined from the momentum representation $\Phi(k)$. Although the uncertainty principle was first discovered by Heisenberg as he developed an abstract and highly mathematical theory of quantum mechanics based on matrix algebra, the relation is consistent with properties common to all waves.

An eigenfunction with even parity ψ_{even} can be represented in terms of a weighted superposition of cosine functions with a continuous spectrum of values of k. In the most general case, the possible values of k range from zero to infinity and the superposition of the functions is represented by an integral over this range. (An eigenfunction with odd parity is represented by the superposition of sine functions. For the sake of simplicity, only eigenfunctions with even parity are considered here.)

$$\psi(x)_{even} = (1/2\pi)^{1/2} \int_{-\infty}^{+\infty} \Phi(k) \cos(kx) dk \qquad (5.38)$$

The function $\Phi(k)$, the momentum representation of the eigenfunction, thus prescribes the amplitudes of the superposed cosine functions that form $\psi(x)$ as a function of k. This function can be found by evaluating the Fourier transform of the position representation $\psi(x)$.

$$\Phi(k) = (1/2\pi)^{1/2} \int_{-\infty}^{+\infty} \psi(x) \cos(kx) dx \qquad (5.39)$$

In practice, the integral of Eq 5.39 is often quite difficult to evaluate using analytical methods. Instead, variations of the numerical methods of the previous sections can be used to evaluate the integral to determine $\Phi(k)$. The function $\Phi(k)$ can be displayed graphically using the familiar graphic template. In principle, numerical integration over an infinite range of spatial positions requires an infinite number of calculations (and infinite computer time). To avoid this problem, the spatial dimensions of the eigenfunction $\psi(x)$ are restricted to a region of finite length P ranging from $-P/2 < x < P/2$. Because the value of $\psi(x)$ is zero outside this range, the limits of integration can be restricted to the range $P/2 < x < -P/2$.

When numerical methods are used to evaluate the integral of Eq 5.39, the momentum representation of the eigenfunction is expressed in terms of a discrete Fourier transform

5.8 FOURIER TRANSFORMS AND THE UNCERTAINTY PRINCIPLE

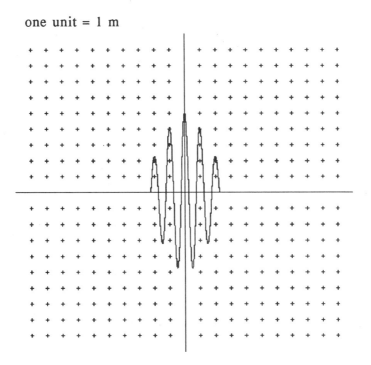

Figure 5.12 This figure illustrates a wave packet drawn using Program 5.5.

in which the integral of Eq 5.38 becomes a sum. Numerical methods for determining the discrete Fourier transform (such as the fast Fourier transform) are based on the definition of the Fourier transform for continuous functions, but are usually applied to perform a spectral analysis of laboratory data sampled over a finite window. For example, the spectral content of an electronic signal can be determined from a digitally stored sample by numerically calculating the discrete Fourier transform. (See R. J. Higgens, *American Journal of Physics*, Vol. 47, p. 766, August 1976.) The numerical determination of the discrete Fourier transform is an important technique for determining spectral content and spectral width of electronic signals and laboratory data. In this section a simple numerical integration is used to generate a graphical representation of the discrete Fourier transform of eigenfunctions of Schrödinger's equation.

The eigenfunction $\psi(x)$ being considered is graphed relative to the graphic template using Program 5.5. The function being plotted is defined in line number 155. The length P of the function is specified in terms of its wavelength in line number 170. The value of the function is calculated as the for-next loop beginning at line number 1010 is executed. The figure is scaled and plotted by execution of the subroutine beginning in line number 3000.

```
 90 REM               *****  Program 5.5  *****
 95 REM     Wave Packets (Position Representation)
100 REM
          *****  set up graphics characteristics  *****

110 SCREEN 2 : CLS : XO = 320 : YO = 100 : SX = 1.5 : SY = SX/2.25
150 REM
          *****  specify initial conditions  *****

155 DEF FNSI(X) = 5*EXP(-X*X/5)*COS(6.28*X/L):REM define eigenfunction SI(x)
156 REM SI(x) must be an even function
160 L = 1 :REM wavelength in meters
165 N = 4.5 :REM number of waves in wave packet
170 P = L*N : REM length of wave packet in meters
180 DX = .01
190 DK = .2
300 REM     *****  set up screen display  *****

310 Y1 = 0 : REM draw horizontal axis
320 FOR X1 = -110 TO 110 STEP 2
330 XS = XO + SX*X1 : YS = YO - SY*Y1 : PSET (XS,YS)
340 NEXT X1
350 X1 = 0 : REM draw vertical axis
360 FOR Y1 = -100 TO 100 STEP 1.5
370 XS = XO + SX*X1 : YS = YO - SY*Y1 : PSET (XS,YS)
380 NEXT Y1
390 REM draw coordinate grid
400 FOR X1 = -100 TO 100 STEP 10
410 FOR Y1 = -90 TO 90 STEP 10
420 XS = XO + SX*X1 : YS = YO - SY*Y1 : PSET (XS,YS)
430 NEXT Y1
440 NEXT X1
450 SC = 1 : REM scale for screen grid in meters
460 SX = 10*SX/SC : SY = SX/2.25
470 LOCATE 1,55 : PRINT "one unit =";SC;"m"
1000 REM
          *****  calculations and plotting  *****

1010 FOR X = -P/2 TO P/2 STEP DX
1020 Y = FNSI(X)
1030 GOSUB 3000
1050 NEXT X
1100 END
2990 REM     *****  plotting subroutine  *****
3000 XS = XO + SX*X : YS = YO - SY*Y : PSET (XS,YS)
3010 RETURN
```

5.8 FOURIER TRANSFORMS AND THE UNCERTAINTY PRINCIPLE

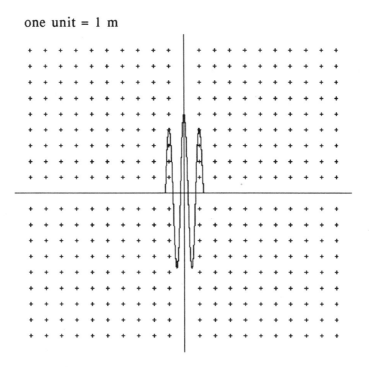

Figure 5.13 This figure illustrates a wave packet drawn using Program 5.5. This wave packet is more localized than the wave packet shown in Figure 5.12.

Figure 5.12 displays the results of Program 5.5. To produce Figure 5.13, only the spatial range L of the eigenfunction was reduced. In order to represent realistic eigenfunctions, the values of L were chosen so that $\psi(x)$ equals zero at the limits of the range. These functions thus correspond to solutions of Schrödinger's equation for the special case of a particle bound inside an infinite potential well.

The discrete Fourier transform is found by evaluating Eq 5.39. Program 5.6 evaluates the integral for the eigenfunction shown in Figure 5.12. The numerical integration is performed by execution of the for-next loop beginning at line number 1030. Successive values of k are specified in the for-next loop beginning at line number 1010. The numerical integration is performed for each of the successive values of k. The results of the integration are scaled and plotted by execution of the subroutine beginning at line number 3000.

```
 90 REM           *****  Program 5.6  *****
 95 REM Fourier Transforms (Momentum Representation)
100 REM
         *****  set up graphics characteristics  *****

110 SCREEN 2 : CLS : XO = 320 : YO = 100 : SX = 1.5 : SY = SX/2.25
```

92   5 · SCHRÖDINGER'S EQUATION

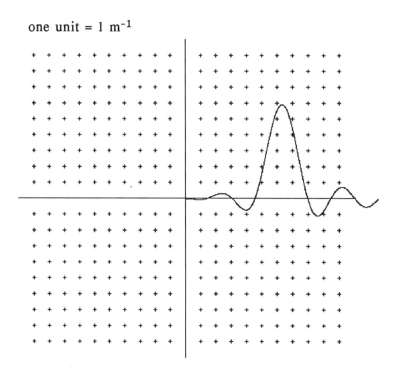

Figure 5.14 This figure depicts the Fourier transform of a wave packet for which the wavelength is one unit and the length of the wave packet is four units.

```
150 REM
              *****  specify initial conditions  *****

155 DEF FNSI(X) = 5*EXP(-X*X/10)*COS(6.28*X/L):REM define
eigenfunction SI(x)
156 REM SI(x) must be an even function
160 L = 1 :REM wavelength in meters
165 N = 4 :REM number of waves in wave packet
170 P = L*N : REM length of wave packet in meters
180 DX = .01
190 DK = .2
300 REM
              *****  set up screen display  *****

310 Y1 = 0 : REM draw horizontal axis
320 FOR X1 = -110 TO 110 STEP 2
330 XS = XO + SX*X1 : YS = YO - SY*Y1 : PSET (XS,YS)
340 NEXT X1
350 X1 = 0 : REM draw vertical axis
360 FOR Y1 = -100 TO 100 STEP 1.5
370 XS = XO + SX*X1 : YS = YO - SY*Y1 : PSET (XS,YS)
380 NEXT Y1
390 REM draw coordinate grid
```

5.8 FOURIER TRANSFORMS AND THE UNCERTAINTY PRINCIPLE

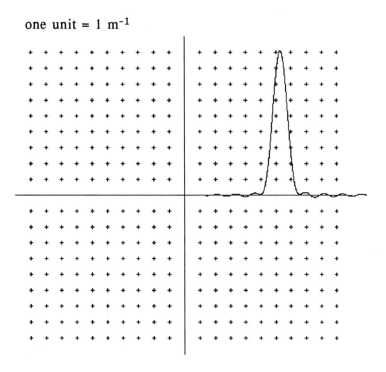

Figure 5.15 For this figure, the length of the wave packet is increased to ten units. Because the wave packet is less localized, the Fourier transform is more localized.

```
400  FOR X1 = -100 TO 100 STEP 10
410  FOR Y1 = -90 TO 90 STEP 10
420  XS = XO + SX*X1 : YS = YO - SY*Y1 : PSET (XS,YS)
430  NEXT Y1
440  NEXT X1
450  SC = 1 : REM scale for screen grid in meters
460  SX = 10*SX/SC : SY = SX/2.25
470  LOCATE 1,55 : PRINT "one unit =";SC;"(1/m)"

1000 REM
             ***** calculations and plotting *****

1010 FOR K = 0 TO 12.5 STEP DK
1020 AK = 0
1030 FOR X = -P/2 TO P/2 STEP DX
1040 AK = AK + FNSI(X)*COS(K*X)*DX
1050 NEXT X
1060 GOSUB 3000
1070 NEXT K
1100 END
2990 REM    ***** plotting subroutine *****
3000 XS = XO + SX*K : YS = YO - SY*AK : PSET (XS,YS)
3010 RETURN
```

The Fourier transform of the function shown in Figure 5.12 is shown in Figure 5.14. The peak in the Fourier transform is centered at $k = 6.28$ m^{-1}. Because $k = 2\pi/L$, this value of k corresponds to wavelength of 1 m, the wavelength of the wave packet shown in Figure 5.12. (See Exercise 5.20.)

Figure 5.14 represents the Fourier transform of a wave packet 4 m in length. This length is the spatial extent of the wave packet and thus represents the uncertainty in position dx. The width of the central peak of the Fourier transform is a measure of the uncertainty dk. The width of the peak in Figure 5.14 is 3 m^{-1}. Thus the product of uncertainties is

$$dxdp = (4)(3) = 12 \tag{5.40}$$

This value is consistent with the predictions of the uncertainty principle as stated in Eq 5.37.

If the length of the wave packet is increased, the uncertainty in position is increased. However, as seen in Figure 5.15, the width of the peak of the Fourier transform is decreased representing an decreased uncertainty in k.

Other eigenfunctions can be studied using this technique. (See Exercise 5.21.) In general, the more localized the eigenfunction $\psi(x)$, the broader the peak of $\Phi(k)$.

Exercises

5.1 a. Using Program 5.1, make measurements of the figure on the computer screen to determine the wavelength of the eigenfunction of the electron bound in the potential well.
b. Compare the measured value of the wavelength with the predictions of the deBroglie postulate $p = \hbar/L$, where $p = [2m(E - V)]^{1/2}$. The total energy E of the bound electron is specified in line number 180 and the potential energy V is zero in the region of the potential well.

5.2 By changing the value of the energy E specified in line number 180 in Program 5.1, determine the ground state of the system.

5.3 a. Modify the initial values of ψ and of $d\psi/dx$ specified in line number 190 and line number 200 in Program 5.1 to produce an eigenfunction with odd symmetry.
b. Increase the value of E determined in Exercise 5.2 to find the first excited state of the system. (The first excited state has odd parity.)
c. Alter the initial value of $d\psi/dx$ to verify that the amplitude of the eigenfunction pictured on the screen is determined by the value of this quantity.

EXERCISES

5.4 a. The energy eigenvalues corresponding to bound states is affected by the depth of the potential well. Alter line number 1020 in Program 5.1 to double the depth of the potential well.

b. Find the energy eigenvalue for the first excited state by changing the value of E found in Exercise 5.3. Does the value of the energy eigenvalue increase or decrease as the depth of the potential well is made greater?

c. Again double the depth of the well and find the energy eigenvalue.

d. Compare the energy eigenvalues for the first excited state of the electron bound in a potential well of finite depth with the results obtained for a particle bound in a potential well of infinite depth

$$E_n = n^2\pi^2\hbar^2/2mL^2 \qquad n = 1, 2, 3 \ldots \qquad (5.41)$$

5.5 a. The energy of a bound state is also affected by the width of the potential well. Alter line number 1010 and line number 1110 of Program 5.1 to reduce the width of the well to one-half of its original width.

b. Reduce the energy of the bound electron to determine the energy eigenvalue of the first excited state. Compare this result with the value found in Exercise 5.3(b).

5.6 Instead of graphing the eigenfunction ψ, calculate and graph the probability density $P(x)$. Calculate the value of a new variable P which is equal to ψ^2. Alter the graphic subroutine in Program 5.1 to plot the value of this variable.

5.7 a. What is the maximum value of the energy E in Program 5.1 for which the electron is bound to the potential well?

b. Because the depth of the well is limited, only a finite number of bound states exist. Determine the number of bound states which are possible for this potential well. Set $E = V$ and reduce E until an eigenvalue is determined. Be sure to check both states with even parity and with odd parity. The number of peaks in the eigenfunction corresponding to the highest energy of a bound state equals the quantum number n and is equal to the number of bound states for the potential well.

5.8 a. Extend the results of Program 5.2 to graph the eigenfunction in the $-x$ direction from the origin. Alter line number 1010 to produce a for-next loop for which x varies from 0 to -20×10^{-10} m in steps of $-dx$.

b. Add this loop to the program to verify that the program produces results with either even parity or odd parity depending on the initial conditions specified in line number 190 and line number 200.

c. Verify that the amplitude of the eigenfunction is determined by the initial value of ψ when the eigenfunction has even parity.

d. Verify that the amplitude of the eigenfunction is determined by the initial value of the $d\psi/dx$ when the eigenfunction has odd parity.

5.9 a. Verify that the energy eigenvalue of the ground state of the electron bound in the harmonic oscillator potential is equal to $\hbar\omega/2$ where $\omega = (k/m)^{1/2}$. (Hint: the ground state has even parity.)

b. Verify that the energy eigenvalues for successive levels is

$$E_n = (n + 1/2)\hbar\omega \qquad n = 0, 1, 2, \ldots \tag{5.42}$$

5.10 a. In accordance with the Correspondence Principle postulated by Niels Bohr in 1924, the behavior of classical systems should be in agreement with the predictions of quantum mechanics for large values of the quantum number n. This concept relates the probability interpretation of quantum mechanics with the results of classical mechanics. Using Eq 5.42, estimate the energy of the energy eigenvalue corresponding to the sixteenth allowed level of the harmonic oscillator potential. Hint: the graphic scale factors must be reduced and the range of the for-next loop beginning in line number 1010 in Program 5.2 must be extended in order graph this result on the computer screen.

b. For an oscillating system, the highest probability density for a classical system occurs in the region where the velocity is lowest as the particle spends more time in this region. Use this idea and the correspondence principle to explain why the amplitude of the eigenfunction is highest in the region near the edge of the potential well.

5.11 a. Alter Program A.2 to plot the potential function

$$V = -kd^2 \frac{x^2 + d^2}{x^4 + 8d^4} \tag{5.43}$$

Scale the graphic template so that one unit along the horizontal axis represents 1×10^{-10} m. Set the value of d equal to 6×10^{-10} m, and set the value of k equal to $60 \times 1.6 \times 10^{-19}$.

b. Alter Program 5.2 to evaluate Schrödinger's equation and graph the eigenfunctions corresponding to the lowest three energy eigenvalues for this potential function. Set the value of k equal to $60 \times 1.6 \times 10^{-19}$.

5.12 The one-dimensional hydrogen atom consists of an electron moving in a one-dimensional potential $-ke^2/|x|$. Because this potential is infinite at the origin ($x = 0$), consider a slightly different potential $-ke^2/(a + |x|)$.

a. Alter Program A.2 to plot the potential function.

b. Choose values so that $ke^2 = 2 \times 10^{-28}$ Jm and $a = 0.1 \times 10^{-10}$ m. Alter Program 5.2 to evaluate Schrödinger's equation and graph the eigenfunctions corresponding to the lowest three energy eigenvalues for this potential. (See R. Louden, *American Journal of Physics*, p. 649, 1959.)

5.13 Using the technique described in Exercise 5.1, verify that the wavelength of the free particle is consistent with deBroglie's postulate.

5.14 a. Alter Program 5.3 to produce a potential step to create Figure 5.8.

b. Alter Program 5.3 to produce a potential step to create Figure 5.9.

EXERCISES

5.15 a. Using the results of Exercise 5.12(a), measure P_{max} and P_{min}, the maximum and minimum height of the probability density from the figure on the computer screen.
b. Using these results, calculate R, the reflection coefficient for this potential step.
c. Increase the height of the step and measure the change in the reflection coefficient R.

5.16 a. Using the results of Program 5.4, measure P_{max} and P_{min}, the maximum and minimum height of the probability density from the figure on the computer screen.
b. Using these results, calculate R, the reflection coefficient for this potential barrier.
c. Increase the height of the barrier and measure the change in the reflection coefficient R.

5.17 a. The reflection coefficient is also affected by the thickness of the barrier. Double the thickness of the barrier in Program 5.4 and determine the value of the reflection coefficient.
b. The reflection coefficient is also affected by the shape of the potential barrier. Instead of specifying a constant value for the potential in the region of the barrier, replace line number 1120 in Program 5.4 with the program line below. This alteration changes the barrier from a constant potential to a variable potential.

```
1120   V = 20*(1.6*10^-19)*EXP(-X*X/W)  : REM Truncated
Gaussian potential with a maximum height of 20 eV
```

5.18 Particles with energy greater than the barrier height can also be reflected from the barrier. Increase the energy of the incident particle to a value greater than the barrier height and measure the reflection coefficient for this system.

5.19 Using Program 5.4, determine the transmission coefficient for a series of energies from zero to a value three times greater than the barrier height. Plot a graph of the values as function of energy. This graph relates the probability of transmission past the barrier as a function of the energy of the incident particle.

5.20 a. Alter Program 5.5 to increase L, the wavelength of the wave packet.
b. Using Program 5.6, verify that the peak in the discrete Fourier transform is centered at a value of k corresponding to $2\pi/L$.

5.21 a. Alter line number 155 of Program 5.5 as shown below to plot a Gaussian function.

```
155 DEF FNSI(X) = 5*EXP(-X*X/5)
```

b. Using Program 5.6, determine the discrete Fourier transform for this function. The Fourier transform in this case has the same form as the original function.

Appendix A

COMPUTER GRAPHICS

A.1 Introduction

Figures created by computers can display the results of calculations. However, with a little imagination and a sound understanding of graphical methods, figures can do more than merely display results. Well-constructed and thoroughly understood graphics can create understanding. When applied to physical systems, graphic displays can relate the behavior of physical systems to the principles being applied. Before students can gain the benefits of this powerful method of creating understanding, a few useful skills must be developed.

The purpose of this appendix is to develop a base of common experience before beginning to use computers in the study of modern physics. Techniques used to create graphic displays and examples of the use of graphical techniques in classical physics are included in this appendix. Students who plan to use computers or programming languages other than those used in this text may wish to refer to Appendix B to see how to alter the programs for use with other computers. If none of the systems described in Appendix B are to be used, students can interpret the function of the graphics operations described and can then determine the syntax required to accomplish the operation with their computer and programming language.

A.2 Screen Concepts

When used to create graphic displays, positions on the screen of the computer monitor are labelled in terms of cartesian coordinates. Each position (x_s, y_s) of the coordinate system is designated by integer values of x_s and y_s and is associated with a picture element (pixel). The subscript s denotes coordinate positions on the computer screen. Images are created by controlling arrays of pixels.

The orientation and proportions of this coordinate system are not the same as those of standard cartesian coordinate systems. The upper left corner of the screen is ordinarily the origin of the coordinate system: the point (0,0). The positions of pixels located increasingly further to the right of the origin are designated by increasing values of the x_s coordinate. For screen 2 of Microsoft BASICA, the graphic system used in the programs of this text, the position of the pixel located at the upper right corner of the screen is (639,0).

The positions of pixels located increasingly further below the origin are designated by increasing values of the y_s coordinate. In the programs of this text, the lower left corner of the screen is designated (0,199) and the lower right corner of the screen is thus (639,199). All of the pixels of the screen are similarly labelled. For example, the pixel nearest the center of the screen is (320,100).

The precision of the images which are formed by controlling this array of pixels is determined by the number of visibly distinct pixels which can be displayed on the screen. The larger the number of pixels, the greater the precision of the images formed on the screen. The screen described above consists of 640 by 200 positions at which a pixel can be displayed.

In addition to precision and location, another important characteristic of pixels is aspect ratio—the ratio of height to width. The aspect ratio of pixels is not standardized and varies with types of computers, programming languages, and even monitors. For the programs of this text, the aspect ratio of pixels is assumed to be 2.25. (Not surprisingly, this is the aspect ratio of the pixels that appear on the author's monitor.) A "square" with dimensions of 20 pixels by 20 pixels when drawn on this screen forms a rectangle having a height 2.25 times greater than its width. A "rectangle" 45 pixels wide and 20 pixels high appears on the screen as a square. A working value of the actual aspect ratio for a particular system can be determined by trial and error. (See Exercise A.1.) When a square appears on the screen, the aspect ratio of the pixels of the screen equals the width of the square in pixels divided by the height of the square in pixels.

A.3 Graphics Operations

In order to minimize the technical knowledge required to use this text, computer programs incorporate only three distinct graphics operations: point plotting, cursor positioning, and text printing. The syntax used in the programs is specific to Microsoft BASICA. A description of the function and syntax of each operation is given in this section. Students who have mastered these operations can add other useful graphics operations such as line drawing, circle drawing, and color control to enhance the appearance and speed of the programs if desired.

1. Point plotting (Display a single pixel)

When creating figures on the screen, it is often necessary to display a single pixel at a specified location. To display the pixel at the position (50,50), the statement PSET(50,50) is executed. If the position of the pixel is specified by variables XS and YS defined earlier in the program, the statement becomes PSET(XS,YS).

A line can be drawn on the screen by repeated execution of the point-plotting statement. When incorporated into a graphics program, the following program line draws a straight line corresponding to the equation $y_s = 2x_s$ for the values of x_s from 0 to 50 by executing the point-plotting statement for each integer value of x_s.

```
100 FOR XS = 0 TO 50: YS=2*XS:PSET(XS,YS):NEXT XS
```

2. Positioning the text cursor

In order to control the position at which text is displayed on the screen of the computer monitor, the text cursor must be placed at the desired position on the screen before text is printed. To place the cursor at the position (X,Y), the statement LOCATE Y,X is executed. It is important to note that the values of X and Y used in this statement refer to coordinates of the text screen rather than coordinates of the graphics screen. The value of X specifies the horizontal position (called a text column) and ranges from 1 to 80. The value of Y specifies the vertical position (called a text row) and ranges from 1 to 25. Execution of the statement LOCATE 1,1 places the cursor at the upper left corner of the screen. The cursor is placed at the bottom right corner of the screen by execution of the statement LOCATE 80,25.

Text is thus positioned relative to a very low precision screen with a resolution of 80 by 25 positions and with an aspect ratio equal to that of the graphics screen (2.25). All screen positions determined relative to the graphics screen must be divided by a factor of 8 when used with the LOCATE statement to position text on the screen.

3. Printing text

After the cursor is positioned, text is printed on the screen using the familiar BASIC language PRINT statement. The first character of the text being printed is placed at the current location of the text cursor. When incorporated into a program, execution of the following program line prints with the first letter of the text placed at the center of the screen.

```
100 LOCATE 12,40 : PRINT "text"
```

A.4 Graphic Template

The coordinate grid shown in Figure A.1 is used as a graphic template in all of the programs of this text. A program which creates the coordinate grid is developed in this section. This program provides an example of graphic programming using the graphics operations described in the previous section. The program allows the grid to be centered

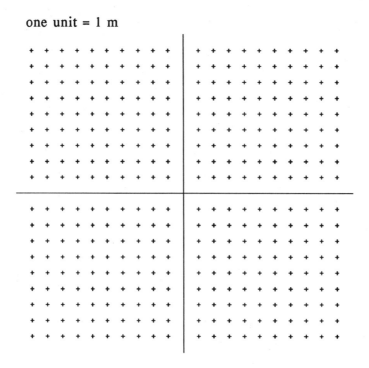

Figure A.1 This graphic template is used for all of the programs of this text.

at any desired screen position. By specifying the scale of the drawing, the size of the grid can also be controlled.

To create the grid, a horizontal line from $x = -11$ to $x = 11$ is drawn 10 units below the top of the grid. A vertical line from $y = -10$ to $y = 10$ is then drawn 11 units from the edge of the grid. A simple program to draw these two lines might be naively written as shown below.

```
100 CLS: SCREEN 2
110 Y = -10
120 FOR X = -11 TO 11
130 PSET(X,Y)
140 NEXT X
150 X = 11
160 FOR Y = -10 TO 10
170 PSET(X,Y)
180 NEXT Y
200 END
```

The graphic screen is cleared and displayed by the execution of line number 100. A horizontal line is created by repeated execution of the point-plotting statement in line number 130. A vertical line is then created point-by-point as line number 170 is executed.

A.4 GRAPHIC TEMPLATE

The figure created by this program, however, is much different than the figure being visualized. Several problems are readily apparent. The lines drawn on the screen are too short, the figure is centered at the top corner of the screen, and the horizontal line is longer than the vertical line.

Solving these problems and creating the desired figure requires finding a way to relate the coordinates (x,y) of the figure to the coordinates (x_s,y_s) of the pixels of the graphic screen. To move the figure to the center of the screen, line number 130 and line number 170 are replaced by the following program lines.

```
130 XS = 320 + X : YS = 100 - Y : PSET(XS,YS)
170 XS = 320 + X : YS = 100 - Y : PSET(XS,YS)
```

With this modification, the program plots each point (x,y) at the screen position (x_s,y_s). In this example, the point $x = 0$, $y = 0$ is plotted at the center of the screen, the position $x_s = 320$, $y_s = 100$. As the value of x is increased, the value of x_s is increased and the point is plotted further to the right. As the value of y is increased, the value of y_s is decreased and the point is plotted at positions higher on the screen.

The program now places the figure on the screen at any specified position. (See Exercise A.2.) A further modification of line number 130 and line number 170 can alter the size and proportions of the figure. To draw a figure of the size shown in Figure A.1, line number 130 and line number 170 are replaced by the following program lines.

```
130 XS = 320 + 15*X : YS = 100 - (15/2.25)*Y: PSET(XS,YS)
170 XS = 320 + 15*X : YS = 100 - (15/2.25)*Y: PSET(XS,YS)
```

With this modification, scale factors are introduced to control the size of the figure. (See Exercise A.2.) In order to create a figure with accurate proportions, the scale factor associated with the vertical screen position y_s is divided by the aspect ratio of the screen (2.25 in this example). The program now produces a horizontal row of points and a vertical row of points. The points located on the horizontal row are 15 pixels apart. Points on the vertical row are located about (15/2.25) pixels apart. The exact distance of separation varies slightly because the locations of pixels correspond to integer values of x_s and y_s. To change the rows of points into continuous lines, line number 120 and line number 160 are replaced by the following program lines.

```
120 FOR X = -11 TO 11 STEP .1
160 FOR Y = -10 TO 10 STEP .1
```

With these changes the program now draws the horizontal and vertical axes of the figure. In general, screen positions x_s and y_s corresponding to values of x and y are determined by equations of the general form of the statements in line number 130 and line number 170.

$$XS = HO + SH*X \tag{A.1}$$

$$YS = VO - SV*Y \tag{A.2}$$

The variables XS and YS represent screen positions in all of the programs of this text. The variables HO and VO specify the screen coordinates of the position corresponding to the point $x = 0$, $y = 0$. SH specifies the horizontal scale factor and SV specifies the vertical scale factor of the figure. The values of SH and SV are selected to produce figures of the desired size and proportions. (See Exercise A.2.) For the programs of this text, the value of SV, the scale factor for vertical positions, is equal to SH divided by the aspect ratio of the pixels of the computer monitor ($SV = SH/2.25$).

Program A.1 below incorporates the ideas just developed to draw the graphic template used in the programs of this text. All of the graphic programs of this text incorporate the graphic template. Once this program is saved as a file, it can be used as a starting point for the development of other programs. Program A.1 is incorporated as a template in all of the programs of this text.

```
 90 REM              *****   Program A.1   *****
 95 REM                    Graphic Template
100 REM
        *****   set up graphics characteristics   *****

110 SCREEN 2 : CLS : XO = 320 : YO = 100 : SX = 1.5 : SY = SX/2.25
150 REM
              *****   specify initial conditions   *****

300 REM
              *****   set up screen display   *****

310 Y1 = 0 : REM draw horizontal axis
320 FOR X1 = -110 TO 110 STEP 2
330 XS = XO + SX*X1 : YS = YO - SY*Y1 : PSET (XS,YS)
340 NEXT X1
350 X1 = 0 : REM draw vertical axis
360 FOR Y1 = -100 TO 100 STEP 1.5
370 XS = XO + SX*X1 : YS = YO - SY*Y1 : PSET (XS,YS)
380 NEXT Y1
390 REM draw coordinate grid
400 FOR X1 = -100 TO 100 STEP 10
410 FOR Y1 = -90 TO 90 STEP 10
420 XS = XO + SX*X1 : YS = YO - SY*Y1 : PSET (XS,YS)
430 NEXT Y1
440 NEXT X1
450 SC = 1 : REM scale for screen grid in meters
460 SX = 10*SX/SC : SY = SX/2.25
470 LOCATE 1,55 : PRINT "one unit =";SC;"m"
500 END
```

A.5 GRAPHING (CARTESIAN)

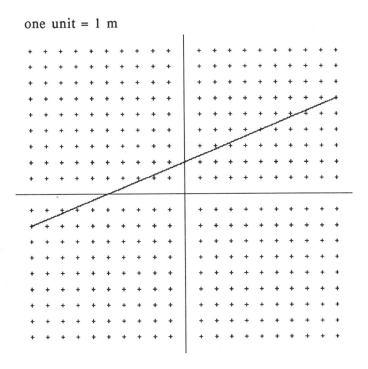

Figure A.2 The straight line $y = 0.5x + 2$ is graphed relative to the graphic template using Program A.2.

A.5 Graphing a Mathematical Function (Cartesian Coordinates)

For a first application, the graphic template is used as a coordinate system to graph algebraic functions expressed in cartesian coordinates. Program A.2 was developed by extending Program A.1 to plot the function $y = mx + b$. The function represents a straight line with slope m and y-intercept b. In the example program, $m = 0.5$ and $b = 2$.

The slope M and intercept B of the straight line to be plotted are specified in line number 160 and line number 170, respectively. The values of y are calculated in line number 1020 for values of x specified as the for-next loop is executed. The GOSUB statement of line number 1030 directs the program to the subroutine at line number 2010. Scaling the function to the graphic template and plotting the points on the computer screen is handled in line number 2010 of this subroutine. The figure created by execution of this program is shown in Figure A.2.

Other functions can be plotted by altering the function defined in line number 1020. For example, Program A.2 can easily be modified to plot the function $y = Ax^2 + Bx + C$. The results of such a change are illustrated in Figure A.3.

APPENDIX A

```
 90 REM              *****  Program A.2  *****
 95 REM           Graphing (Cartesian Coordinates)
100 REM
         *****  set up graphics characteristics  *****

110 SCREEN 2 : CLS : XO = 320 : YO = 100 : SX = 1.5 : SY = SX/2.25
150 REM
             *****  specify initial conditions  *****

160 M = .4 : REM slope of line
170 B = 2 : REM y intercept of line

300 REM
                *****  set up screen display  *****

310 Y1 = 0 : REM draw horizontal axis
320 FOR X1 = -110 TO 110 STEP 2
330 XS = XO + SX*X1 : YS = YO - SY*Y1 : PSET (XS,YS)
340 NEXT X1
350 X1 = 0 : REM draw vertical axis
360 FOR Y1 = -100 TO 100 STEP 1.5
370 XS = XO + SX*X1 : YS = YO - SY*Y1 : PSET (XS,YS)
380 NEXT Y1
390 REM draw coordinate grid
400 FOR X1 = -100 TO 100 STEP 10
410 FOR Y1 = -90 TO 90 STEP 10
420 XS = XO + SX*X1 : YS = YO - SY*Y1 : PSET (XS,YS)
430 NEXT Y1
440 NEXT X1
450 SC = 1 : REM scale of screen grid in meters
460 SX = 10*SX/SC : SY = SX/2.25
470 LOCATE 1,55 : PRINT "one unit = ";SC;"m"
1000 REM
             *****  calculations and plotting  *****

1010 FOR X1 = -10 TO 10 STEP .1
1020 Y1 = M*X1 + B
1030 GOSUB 2010
1040 NEXT X1
1050 END
2000 REM
                *****  Plotting subroutine  *****

2010 XS = XO + SX*X1 : YS = YO - SY*Y1 : PSET (XS,YS)
2020 RETURN
```

A.5 GRAPHING (CARTESIAN)

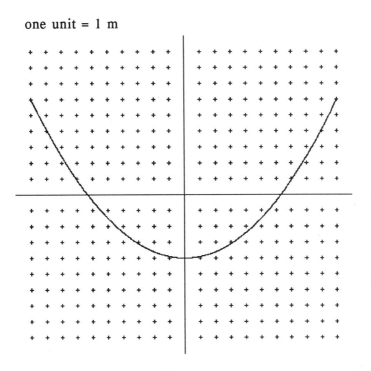

Figure A.3 A parabola is graphed relative to the graphic template using Program A.2.

A.6 Graphing a Mathematical Function (Polar Coordinates)

Polar functions of the form $r = r(\theta)$ are plotted by converting the polar coordinates (r,θ) to cartesian coordinates (x,y) using Eq A.1 and Eq A.2.

$$x = r \cos \theta \quad (A.1)$$

$$y = r \sin \theta \quad (A.2)$$

Program A.3 provides an example of this technique. In this program, the polar form of the equation for a circle (r = constant) provides an example of a polar function.

The radius of the circle (represented by RO) is specified in line number 160. A for-next loop beginning at line number 1010 is used to specify values of the angular position θ (represented by TH) at which the radial position r is to be calculated. The value of R is determined in line number 1020. Equations A.1 and A.2 are applied in line number 1030 and line number 1040. As in Program A.2, the screen coordinates corresponding to the values of x and y are determined in line number 2010.

Other polar functions are plotted in a similar manner. For example, Program A.3 is modified to plot a cardioid function $r = r_0 \sin \theta/2$ shown in Figure A.5 by replacing line number 1020 with the program line shown below. Line number 1020 can be altered to plot any desired polar function.

```
1020   R = RO*SIN(TH/2)
```

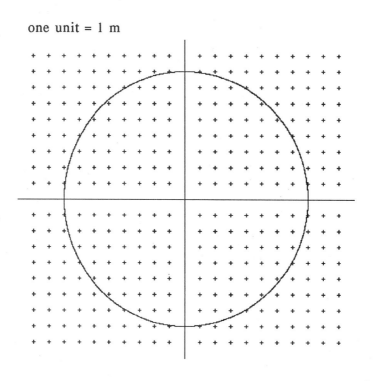

Figure A.4 A circle is drawn when Program A.3 is executed. When the altered program is executed, the figure shown in Figure A.5 is created instead of a circle.

```
 90 REM              *****   Program A.3   *****
 95 REM            Graphing (Polar Coordinates)
100 REM
        *****  set up graphics characteristics  *****

110 SCREEN 2 : CLS : XO = 320 : YO = 100 : SX = 1.5 : SY = SX/2.25
150 REM
             *****  specify initial conditions  *****
160 R0 = 8
300 REM
                *****  set up screen display  *****
310 Y1 = 0 : REM draw horizontal axis
320 FOR X1 = -110 TO 110 STEP 2
330 XS = XO + SX*X1 : YS = YO - SY*Y1 : PSET (XS,YS)
340 NEXT X1
350 X1 = 0 : REM draw vertical axis
360 FOR Y1 = -100 TO 100 STEP 1.5
370 XS = XO + SX*X1 : YS = YO - SY*Y1 : PSET (XS,YS)
380 NEXT Y1
390 REM draw coordinate grid
400 FOR X1 = -100 TO 100 STEP 10
```

A.6 GRAPHING (POLAR)

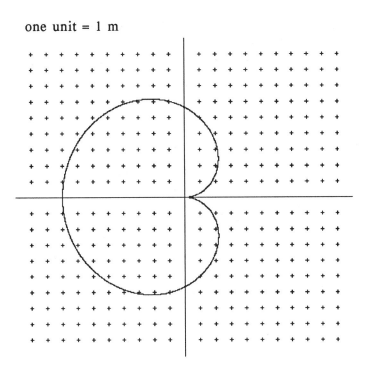

Figure A.5 Program A.3 can be modified to plot a cardioid function.

```
410 FOR Y1 = -90 TO 90 STEP 10
420 XS = XO + SX*X1 : YS = YO - SY*Y1 : PSET (XS,YS)
430 NEXT Y1
440 NEXT X1
450 SC = 1 : REM scale for screen grid in meters
460 SX = 10*SX/SC : SY = SX/2.25
470 LOCATE 1,55 : PRINT "one unit =";SC;"m"
1000  REM
            *****  calculations and plotting  *****

1010 FOR TH = 0 TO 6.28 STEP .01
1020 R = R0
1030 X1 = R*COS(TH)
1040 Y1 = R*SIN(TH)
1050 GOSUB 2000
1060 NEXT TH
1100 END
2000 REM
            ***** plotting subroutine *****

2010 XS = XO + SX*X1 : YS = YO - SY*Y1 : PSET (XS,YS)
2020 RETURN
```

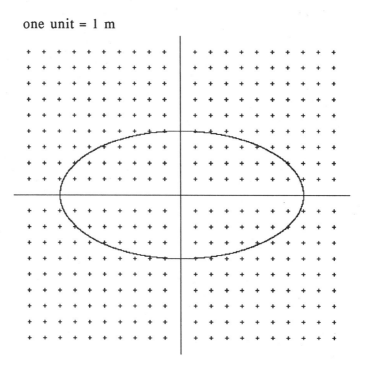

Figure A.6 An ellipse can be drawn using parametric Equations A.3 and A.4.

In addition to polar functions of the form $r = r(\theta)$, polar functions can be written as sets of parametric equations. Equations A.3 and A.4 are the parametric equations describing an ellipse whose axes lie along the coordinate axes.

$$x = C \cos \theta \tag{A.3}$$

$$y = D \sin \theta \tag{A.4}$$

The values of C and D determine the eccentricity of the ellipse. If C and D are equal, the figure described by the equations is a circle of radius C. The following program lines alter Program A.3 to plot an ellipse based on Eq A.3 and Eq A.4. The results of this change are shown in Figure A.6.

```
160  C = 8 : D = 4
1030 X = C*COS(TH)
1040 Y = D*SIN(TH)
```

A.7 Kinematics in Two Dimensions

For a projectile fired with an initial velocity v_0 at an angle θ relative to the horizontal axis, the horizontal component of the initial velocity is $v_x = v_0 \cos \theta$. The initial value of the vertical component of the velocity of the projectile is $v_y = v_0 \sin \theta$. The motion of a projectile is represented by Eq A.5 and Eq A.6 where g is the value of the acceleration due to gravity.

A.7 KINEMATICS IN TWO DIMENSIONS

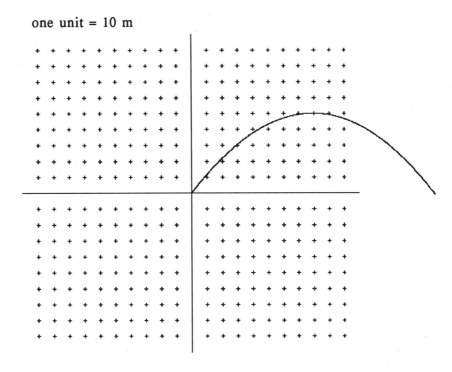

Figure A.7 The motion of a projectile is illustrated using Program A.4.

$$x = v_x t \qquad (A.5)$$

$$y = v_y t + gt^2/2 \qquad (A.6)$$

This simple motion can be simulated on the computer screen by calculating the coordinates (x,y) of the projectile for successive time increments and plotting these coordinate locations.

In Program A.4, v_x and v_y are represented by the variables VX and VY whose values are determined in line number 185. As the for-next loop beginning in line number 1010 is executed, successive positions of the projectile are calculated in line number 1020 and line number 1030 using Eq A.5 and Eq A.6, respectively. As in earlier programs, the coordinate locations are scaled and plotted relative to the graphic template as line number 2010 of the plotting subroutine is executed. Figure A.7 is created using Program A.4. To create Figure A.8, the program was altered to plot the paths of projectiles fired at a series of angles.

```
 90 REM              *****  Program A.4  *****
 95 REM                    Projectile Motion
100 REM
             *****  set up graphics characteristics  *****

110 SCREEN 2 : CLS : XO = 320 : YO = 100 : SX = 1.5 : SY = SX/2.25
```

112 APPENDIX A

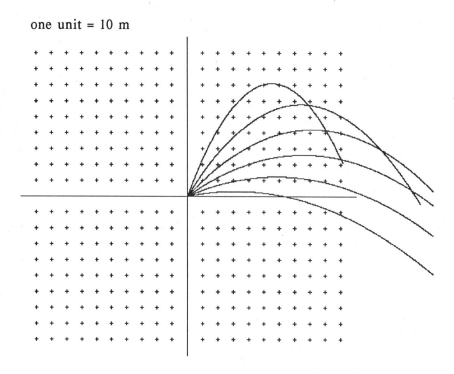

one unit = 10 m

Figure A.8 Projectile motion is illustrated for a succession of firing angles.

```
150 REM
        *****  specify initial conditions  *****

160 V0 = 40  : REM initial velocity of projectile
170 A = .9 : REM angle in radians at which projectile is
fired
180 G = -9.8 : REM acceleration due to gravity
185 VX = V0*COS(A) : VY = V0*SIN(A)
190 DT = .1
300 REM
        *****  set up screen display  *****

310 Y1 = 0 : REM draw horizontal axis
320 FOR X1 = -110 TO 110 STEP 2
330 XS = XO + SX*X1 : YS = YO - SY*Y1 : PSET (XS,YS)
340 NEXT X1
350 X1 = 0 : REM draw vertical axis
360 FOR Y1 = -100 TO 100 STEP 1.5
370 XS = XO + SX*X1 : YS = YO - SY*Y1 : PSET (XS,YS)
380 NEXT Y1
390 REM draw coordinate grid
400 FOR X1 = -100 TO 100 STEP 10
410 FOR Y1 = -90 TO 90 STEP 10
420 XS = XO + SX*X1 : YS = YO - SY*Y1 : PSET (XS,YS)
```

A.7 KINEMATICS IN TWO DIMENSIONS 113

```
430 NEXT Y1
440 NEXT X1
450 SC = 10   : REM scale for screen grid in meters
460 SX = 10*SX/SC : SY = SX/2.25
470 LOCATE 1,55 : PRINT "one unit =";SC;"m"
1000 REM
          *****  calculations and plotting  *****

1010 FOR T = 0 TO 7   STEP DT
1020 X1= VX*T
1030 Y1= VY*T + .5*G*(T^2)
1040 GOSUB 2000
1050 NEXT T
1100 END
2000 REM     *****  plotting subroutine *****
2010 XS = XO + SX*X1: YS = YO - SY*Y1: PSET (XS,YS)
2020 RETURN
```

A.8 Travelling Waves

A continuous sinusoidal travelling wave is characterized by a wavelength L, frequency f, and speed v. The speed of the wave is related to the wavelength and frequency by Eq A.7.

$$v = fL \tag{A.7}$$

A one-dimensional sinusoidal travelling wave of amplitude A moving in the $+x$ direction is described by Eq A.8. A similar travelling wave moving in the $-x$ direction is described by Eq A.9. Here, y represents the height of the wave at a position x.

$$y = A \cos[2\pi(x - vt)/L] \tag{A.8}$$

$$y = A \cos[2\pi(x + vt)/L] \tag{A.9}$$

The appearance of the waves at an instant t can be determined by graphing these functions. In order to study the motion of the wave, graphs can be created for any specified time t. Program A.5 can be used to determine the shape and position of a wave travelling in the $+x$ direction at various times.

The wavelength L, amplitude A, and speed V of the wave are specified in line numbers 160 to 180. A for-next loop beginning at line number 1005 specifies the times for which the wave is to be graphed. A nested for-next loop beginning at line number 1010 specifies the spatial positions x for which the height of the wave is to be calculated and plotted. Execution of the subroutine starting at line number 2010 scales the figure to the graphic template and plots the points on the graphic screen. Results from this program are shown in Figure A.9.

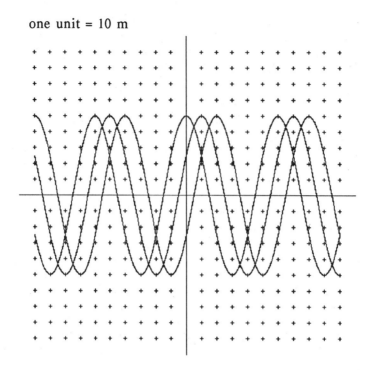

Figure A.9 A travelling wave moving at a speed of 10 m/s is plotted at time intervals of 1 s using Program A.5.

```
 90 REM              *****  Program A.5  *****
 95 REM                  Travelling Waves
100 REM
         *****  set up graphics characteristics  *****

110 SCREEN 2 : CLS : XO = 320 : YO = 100 : SX = 1.5 : SY = SX/2.25
150 REM
              *****  specify initial conditions  *****
160 L = 60
170 A1 = 50
180 V = 10
300 REM
                 *****  set up screen display  *****

310 Y1 = 0 : REM draw horizontal axis
320 FOR X1 = -110 TO 110 STEP 2
330 XS = XO + SX*X1 : YS = YO - SY*Y1 : PSET (XS,YS)
340 NEXT X1
350 X1 = 0 : REM draw vertical axis
360 FOR Y1 = -100 TO 100 STEP 1.5
370 XS = XO + SX*X1 : YS = YO - SY*Y1 : PSET (XS,YS)
380 NEXT Y1
```

A.8 TRAVELLING WAVES

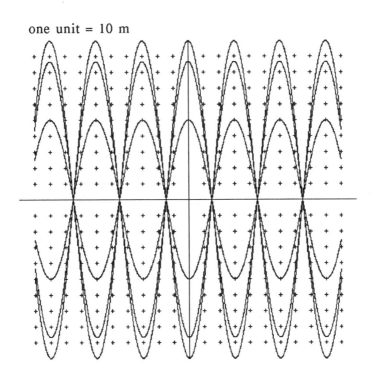

Figure A.10 Program A.5 can be altered to plot a standing wave at time intervals of 1 s.

```
390  REM draw coordinate grid
400  FOR X1 = -100 TO 100 STEP 10
410  FOR Y1 = -90 TO 90 STEP 10
420  XS = XO + SX*X1 : YS = YO - SY*Y1 : PSET (XS,YS)
430  NEXT Y1
440  NEXT X1
450  SC = 10 : REM scale for screen in meters
460  SX = 10*SX/SC : SY = SX/2.25
470  LOCATE 1,55 : PRINT "one unit =";SC;"m"
1000 REM
         *****  calculate values and plot function  *****

1005 FOR T = 0 TO 2 STEP 1
1010 FOR X1 = -100 TO 100
1020 Y1 = A1*COS(6.28*(X1 - V*T)/L)
1030 GOSUB 2000
1040 NEXT X1
1050 NEXT T
1100 END
2000 REM
         ***** plotting subroutine *****

2010 XS = XO + SX*X1 : YS = YO - SY*Y1 : PSET (XS,YS)
2020 RETURN
```

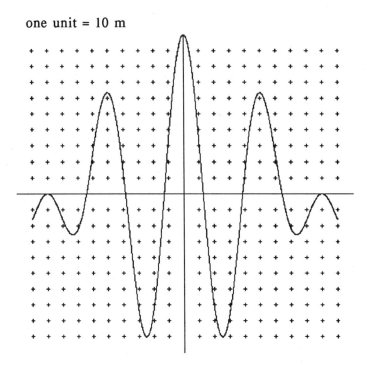

Figure A.11 The motion of the wave group can be studied by plotting the function at a succession of times.

A standing wave results from the superposition of identical travelling waves moving in opposite directions. The equation representing travelling waves results from the addition of Eq A.8 and Eq A.9.

$$y = A \cos(2\pi(x - vt)/L) + A \cos(2\pi(x + vt)/L) \qquad (A.10)$$

Program A.5 can be altered to display standing waves by replacing Eq A.8 in line number 1020 with Eq A.10. The results obtained from execution of Program A.5 with these changes were shown in Figure A.10.

```
1020 Y = A*COS((6.28/L)*(X-V*T)) + A*COS((6.28/L)*(X+V*T))
```

The superposition of two sinusoidal waves travelling in the same direction but with slightly different speeds and wavelengths can be used to study wave groups. By plotting the wave groups at various times t, it is possible to distinguish the group velocity and the phase velocity of the waves. (See Exercise A.12.) Program A.5 can easily be altered to produce these results. In the program lines below, L1 and V1 represent the wavelength and speed of one of the waves, while L2 and V2 represent the wavelength and speed of a second wave. Line number 1020 is altered to calculate the effects of this change. Results from this program for a single instant t are shown in Figure A.11. A sequence of these figures can be created to explore the movement of the wave groups. (See Exercise A.12.)

EXERCISES

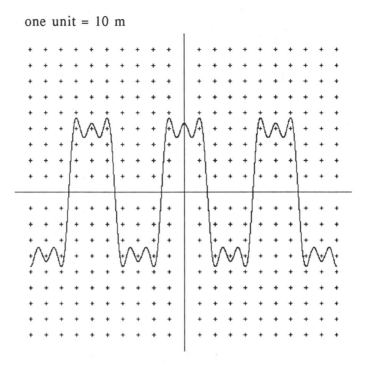

one unit = 10 m

Figure A.12 The first four terms of the Fourier series which synthesize a square wave were plotted to generate this figure.

```
160    L1 = 45: L2 = 60: V1 = 10: V2 = 10
1020   Y =   A*COS((6.28/L1)*(X-V1*T)) +
             A*COS((6.28/L2)*(X-V2*T))
```

Continuous waves of any shape can be synthesized by the addition of an infinite number of terms in a Fourier series. For example, a symmetric square wave results when the following series of terms are added. It is left as an exercise for the student to modify Program A.5 to create wave forms from the addition of terms in a Fourier series. Figure A.12 shows the results of adding the first four terms in the series to required to synthesize a square wave.

Exercises

A.1 The following program draws a horizontal line and then a vertical line each 40 pixels in length centered on the screen.

```
100 SCREEN 2: CLS
110 YS=100:FOR XS = 300 to 340: PSET(XS,YS) : NEXT XS
120 XS=320:FOR YS = 80 TO 120 : PSET(XS,YS) : NEXT YS
130 END
```

a. Measure the length of each line and calculate the aspect ratio (the ratio of height to width) for the pixels of the graphic screen with which you are working.
b. Using the measured values of the aspect ratio, alter the program to produce horizontal and vertical lines of equal length.

A.2 Complete the following exercises by changing the values of the variables HO, VO, SH, and SV in Program A.1.
a. Draw the coordinate grid so that its dimensions are reduced by one-half.
b. Center the reduced scale grid at the pixel (200,100).
c. Draw the grid as large as possible on the screen.

A.3 Complete the following exercises by altering the for-next loops of Program A.1.
a. Eliminate the vertical axis from the grid.
b. Eliminate the grid and retain only the horizontal and vertical axes.
c. Draw the horizontal axis 20 units in length.
d. Make the coordinate grid 20 units wide by 10 units high.

A.4 Modify Program A.2 to plot the following functions.
  a. $y = x$
  b. $y = x^2 - 8$
  c. $y = 4 \sin x$
  d. $y = 4 \sin^2 x$
  e. $y = 4 \exp(-x^2/100) \sin^2 4x$

A.5 Modify Program A.3 to plot the following polar functions.
  a. $r = r_0 \sin(n\theta)$ $n = 1, 2, 3, \ldots$
  b. $r = a\theta$
  c. $r = a/\theta$

A.6 Modify Program A.3 to plot Lissajous curves (Bowditch curves) using the following parametric equations.
$$x = \cos(nt) \quad n = 1, 2, 3, \ldots$$
$$y = \sin(mt) \quad m = 1, 2, 3, \ldots$$

A.7 Modify Program A.3 to draw a family of ellipses for which the horizontal dimensions have integer values ranging from 0 to 16 units.

EXERCISES

A.8 By replacing line number 1010 of Program A.3 with the program line below, the resulting figure produced by the program is a square. (Why?)

```
1010    FOR TH = 0 to 6.28 step 1.57
```

a. Alter the program line to produce a hexagon.
b. Alter the program line to produce an octagon.
c. A "spiral" figure is produced if the program line below is used. (Why?)

```
1010    FOR TH = 0 to 628 step 1.50
```

A.9 Modify Program A.4 to simulate the motion of an object starting from the point (-10,4) and moving across the grid with a constant velocity which has components v_x = 10 m/s, v_y = -4 m/s.

A.10 Modify Program A.4 to simulate the motion of an object undergoing free fall starting at rest from the point (5,3).

A.11 Using Program A.5, measure the speed of the travelling wave drawn on the screen and verify the relation $v = fL$. Graph the wave at two successive times. Measure the distance that the wave moves during the time interval by direct measurement of the position of the wave on the computer screen. To determine the speed of the wave, divide the distance that the wave travels by the time interval between the drawings.

A.12 Plot equations for a wave group at various times and verify the equations for the phase velocity and group velocity of the waves.

A.13 Alter Program A.5 to produce Figure A.11 using the first four terms of the Fourier series of a square wave.

Appendix B

PROGRAM CONVERSIONS

B.1 Introduction

The programs of this text can easily be converted to other programming languages. The principles of Appendix A are general and can be applied with any computer or programming language with graphic capability.

The programs were written for use with the IBM-PC or PS/2 using screen 2 which is available with almost all graphic adaptors. For any of the programs of this text, higher resolution screens can be employed by altering only line number 110 and line number 460. For example, to use screen 11 and increase the graphic resolution to 640 by 480 pixels, the following lines can be substituted.

```
110   SCREEN 11 : CLS : XO = 320 : YO = 240 : SX = 1.5 : SY = SX
460   SX = 10*SX/SC : SY = SX
```

The programs can be used "as is" with Microsoft QuickBASIC. This programming language has extended capability and program execution is many times faster if a mathematics co-processor is used. In addition, scalable graphics "windows" are available and highly structured programs can be created. A variant of this language is quite popular with users of the Apple Macintosh.

With some alteration, the programs can be changed to the True BASIC programming language. As with QuickBASIC, program execution can be much faster, especially if a mathematics co-processor is available. True BASIC programs can be used with either an Apple Macintosh or an IBM-PC with no alteration.

Conversion to other programming languages such as PASCAL requires greater effort. However, the program operations are available in any language, and conversion will prove to be quite simple for programmers even with limited experience. Students with little programming experience will get maximum results with minimum effort using the programs with no alterations with QuickBASIC.

B.2 Applesoft BASIC

In addition to other programming languages, computers other than the IBM-PC can also be used. For example, the programs can easily be converted to run using Applesoft BASIC and the Apple 2e or Apple 2c. The standard screen resolution for these computers is 280 by 192 pixels. In order to convert the programs, the point-plotting command PSET(XS,YS) must be changed to HPLOT(XS,YS). Line number 470 must be omitted in all of the programs because Applesoft BASIC cannot display text on the graphic screen. The following program lines must be substituted. Typing TEXT and pressing RETURN brings the computer back to the text display after graphic programs are run.

```
110 HGR2 : HCOLOR=7 : XO = 140 : YO = 96 : SX = .75 : SY = SX
460 SX = 10*SX : SY = SX
```

B.3 DEC ReGIS

The DEC ReGIS graphics language is used with terminals and microcomputers manufactured by Digital Equipment Corporation (DEC). The system uses specially equipped terminals (such as the VT-240 and VK-100 terminals or Rainbow microcomputers) as interpreters for graphics commands imbedded in output from host minicomputers. These systems can thus respond to output from almost any computer in the context of almost any programming language. The point-plotting command PSET(XS,YS) must be changed to PRINT"P[";XS;",";YS;"]V[]". Line number 470 can be omitted from any of the programs. The following program lines must be substituted.

```
100  PRINT CHR$(27)+"Pp" : PRINT"S(E)" : XO = 320 : YO = 200 : SX = 1.5 : SY = SX
460 SX + 10*SX : SY = SX
```